Logic, Argumentation & Reasoning

Interdisciplinary Perspectives from the Humanities and Social Sciences

Volume 15

Series editor
Shahid Rahman

Logic, Argumentation & Reasoning explores the links between Humanities and the Social Sciences, with theories including, decision and action theory as well as cognitive sciences, economy, sociology, law, logic, and philosophy of sciences. It's two main ambitions are to develop a theoretical framework that will encourage and enable interaction between disciplines as well as to federate the Humanities and Social Sciences around their main contributions to public life: using informed debate, lucid decision-making and action based on reflection.

The series welcomes research from the analytic and continental traditions, putting emphasis on four main focus areas:

- Argumentation models and studies
- Communication, language and techniques of argumentation
- Reception of arguments, persuasion and the impact of power
- Diachronic transformations of argumentative practices

The Series is developed in partnership with the Maison Européenne des Sciences de l'Homme et de la Société (MESHS) at Nord - Pas de Calais and the UMR-STL: 8163 (CNRS).

Proposals should include:

- A short synopsis of the work or the introduction chapter
- The proposed Table of Contents
- The CV of the lead author(s)
- If available: one sample chapter

We aim to make a first decision within 1 month of submission. In case of a positive first decision the work will be provisionally contracted: the final decision about publication will depend upon the result of the anonymous peer review of the complete manuscript. We aim to have the complete work peer-reviewed within 3 months of submission.

The series discourages the submission of manuscripts that contain reprints of previous published material and/or manuscripts that are below 150 pages/85,000 words.

For inquiries and submission of proposals authors can contact the editor-in-chief Shahid Rahman via: shahid.rahman@univ-lille3.fr or managing editor, Juan Redmond at juan.redmond@uv.cl.

More information about this series at http://www.springer.com/series/11547

Jens Erik Fenstad

Structures and Algorithms

Mathematics and the Nature of Knowledge

Springer

Jens Erik Fenstad
Institute of Mathematics
University of Oslo
Oslo
Norway

ISSN 2214-9120 ISSN 2214-9139 (electronic)
Logic, Argumentation & Reasoning
ISBN 978-3-030-10294-4 ISBN 978-3-319-72974-9 (eBook)
https://doi.org/10.1007/978-3-319-72974-9

Softcover re-print of the Hardcover 1st edition 2018

Printed on acid-free paper

This Springer imprint is published by Springer Nature
The registered company is Springer International Publishing AG
The registered company address is: Gewerbestrasse 11, 6330 Cham, Switzerland

Preface

This book is a collection of essays on *what there is* and *how we know*. We are interested in understanding the general nature of human knowledge. Structures and algorithms will be essential tools for this understanding: what we see and how we can make use of what we see. Mathematics will be the link between the two.

But mathematics is not exactly what common man believes; this is the theme of the first essay. When read in the proper way, this essay is an attempt to extend the current art and science of mathematical modeling to all fields of knowledge. The introductory text will be followed by several essays on the "mathematical way", with examples from art, history, language as well as the natural and social sciences.

In the last essay, we return to the question of what there is. We shall argue for a cultural foundation of knowledge. Culture is grounded in man and nature, but exists outside of space and time. This means that numbers, language, and mind are abstract objects in the "collective mind of the species," to use the language of some anthropologists. Structures are what we see. And being abstract cultural objects, they can almost always be seen in many different ways. But certain structures, such as the natural numbers and the basic notions of grammar, seem to have an absolute character. Any theory of knowledge grounded in human culture must explain how this is possible. In the last essay, we present our analysis of this cultural invariance, combing insights from evolutionary theory and current neuroscience.

The book is not a research monograph, nor a textbook. It is a collection of essays offered to any reader who seeks a deeper understanding of knowledge in present-day society. The mathematical and computational sciences have gained in importance, but the understanding of "the mathematical way" is not always as it should be in our society. This is, above all, a challenge to our educational system—from elementary school to research training.

Note on the essays. After an *introductory essay* on mathematics and the nature of knowledge, I have in Part I included five essays on *structure and knowledge* from the *European Review*, the journal of Academia Europaea. They are written for a general audience and illustrate through a number of examples how "the mathematical way" gives insight and understanding. This is a theme that I have developed over many years. As a young postdoc in the early 1960s, I took an active part in a series of lectures arranged by the Norwegian Mathematical Society on the growing importance of mathematical modeling and computations in science and industry; the report I wrote of this event could have been part of this collection of essays. As Chair of the Natural Science Research Council of Norway, I could in the mid-1980s contribute to the financing of the first Cray supercomputer to Norwegian research and the text of my lecture at the opening ceremony could equally have been part of this collection. The same is true of the input on mathematics and computing to the 5th Framework Program of the EU that I developed as Chair of the Standing Committee for Physics and the Engineering Sciences of the European Science Foundation in the mid-1990s. I stressed, in particular, the applicability of "the mathematical way" to language and the human sciences, as well as to the social and medical sciences.

The essays of Part I are followed in Part II by three essays on *language, mind*, and *number*. The first two of these essays are included in order to give a simple example of "the mathematical way" as applied to language and meaning. Grammar, geometry, and mind have been major research interests, starting with a first paper on "Models for Natural Languages" from the 1970s. There is, however, no mathematical way, connecting structures and algorithms, without numbers, either directly as objects or more generally as tools. Understanding knowledge thus means to know what numbers are. This is the topic of the last essay of this collection.

I have tried in this book to navigate around the actual formalism of mathematics, but in the two essays on language and meaning (Chaps. 7 and 8), the reader will find a few formulas and some diagrams to illustrate the general analysis. No other item in this collection, including the last essay on the nature of numbers, presupposes knowledge of the mathematical details of these essays.

The essays have been written over a number of years. This means that what was "recent" at the time, now sometimes needs to be updated. The "mathematical way" and the basic analysis have remained unchanged over time, but some examples related to computing and "big data" need to be revised. The reader will find the necessary updates in the introductory essay.

The essays can be read in almost any order; this means that there is some overlap between the chapters, in particular, in the introductory parts to Chaps. 5 and 7.

Permission to reprint. Chapters 2–9 have previously been published and are here reprinted, sometimes with a few corrections and updates, as follows: (i) Chapters 2, 3, 5, and 6 are reprinted with permission from Cambridge University Press, the publisher of the *European Review*. (ii) Chapter 4 was originally published in *Formative Years of Scholars*, London: Portland Press, and is here reprinted with permission from Cambridge University Press, the publisher of the *European Review*. (iii) Chapter 7 is reprinted with permission from Elsevier, the publisher of the *Annals of Pure and Applied Logic*. (iv) Chapter 8 is reprinted with permission from Springer-Verlag, the current copyright holder of X. Arrazola et al, *Discourse, Interaction and Communication*. (v) Chapter 9 is reprinted with permission from Taylor & Francis, the publisher of *Inquiry*.

Acknowledgements Many people have given good advice and useful comments in the written of this book. I want above all to thank Dagfinn Føllesdal, Øystein Linnebo, Richard Tieszen, and the Springer referee for many helpful and critical comments on several drafts of the text. It is also a pleasure to acknowledge the expert and helpful assistance of my Springer editors in production the book.

Oslo, Norway Jens Erik Fenstad

Contents

Chapter 1
Mathematics and the Nature of Knowledge—An Introductory Essay

Abstract This book is a collection of essays on mathematics and the nature of knowledge. We claim that the mathematical sciences, mathematics, statistics and computing, are almost everywhere. In this introductory essay we present in brief our argument why these sciences are essential for human thought and action. The main body of the text presents further examples from the natural and social sciences, and also from language and art.

Keywords Mathematics—the nature of knowledge · Structures—algorithms Mathematical modeling · Big data—computational feasibility

1.1 What Common Man Believes

The purpose of these essays is to convince common man of the necessity of the mathematical sciences for understanding both the nature of knowledge and how knowledge is applicable in everyday's affairs. Knowledge is grounded in what we "see", in what we observe and what we measure. But knowledge is more than data and observations, knowledge is about patterns and structure in what we "see". Sometimes is what we "see" an end in itself, but if we want to act upon what we "see", we need rules and recipes or, in current usage, algorithms to guide our actions. *Structure* and *algorithms algorithms*will be the key concepts in our quest to understand what knowledge is and how it can be applied. But structure and algorithms must be linked, and *mathematics*, understood in the correct way, is the link between the two.

What common man believes. Mathematics, as learned in school, often appears to the common man as a vast collection of numbers, equations and complex geometrical constructions, to be learnt on command and to be applied with more or less skill in problems of counting and measurements. After years in school common man is satisfied as long as he can control his pay-check and pay his bills. Otherwise mathematics, as far as he remembers the subject from his school days, seems to be largely irrelevant for his daily life. This may, however, be rapidly changing. Information technology in the form of computers, smartphones, robots and other "intelligent" devices plays an increasingly larger role in our lives. But any technology

J. E. Fenstad, *Structures and Algorithms*, Logic, Argumentation & Reasoning 15,
https://doi.org/10.1007/978-3-319-72974-9_1

is grounded in knowledge, and the mathematical sciences, seen as the link between structure and algorithms, are part of every such knowledge base. But is this the mathematics we met at school?

What physicists believe. We first seek advice from the physicists, the tribe behind the modern industrial technology. They will almost certainly point to the names of Galileo, Newton, Einstein and Dirac, and you will understand that for physicists it is equations that rule. Mathematics is, in their view, the language of nature, and the laws of nature are expressed in mathematical equations. And the equations are everything: "The power of equations can seem magical. ... They can take on a power and life of their own, giving birth to consequences that their creator did not expect, cannot control, and may even find repugnant." This is a quote from F. Wilczek's contribution to a book on the great equations of modern science (Wilczek 2002). Wilczek further cites with approval the following confession once made by Dirac and which is quite characteristic of Dirac's approach to science: "A great deal of my work is just playing with equations ..."

For Wilczek it is equations that express the true nature of the scientific method. He describes this approach as the method of 'experimental logic': "In experimental logic, one formulates hypotheses in equations, and experiments with those equations. That is, one tries to improve the equations from the point of view of beauty and consistency, and then checks whether the 'improved' equations elucidate some feature of nature." No one will deny the genius of Dirac, but is "playing with equations" the full story about mathematics and nature?

Structures and patterns. There are many, even among the physicists, who disagree with Dirac. In his contribution to the book *Seeing Further*, celebrating the 350th anniversary of the Royal Society of London, the physicist J. D. Barrow forcefully argues how structures and patterns come before equations: "Mathematics is simply the catalogue of all possible patterns. Some of those patterns are especially attractive and are studied for decoration, others are patterns in time or in chains of logic. Some are described solely in abstract terms, while others can be drawn on paper or carved in stone. Viewed in this way, it is inevitable that the world is described by mathematics", Barrow (2010).

1.2 The Mathematical Way

Following Barrow we shall argue that to understand the mathematical sciences and how they are applied, is to understand the family of concepts: *structures*, *syntax* and *algorithms*. This is clearly understood in many fields of science today, let numerical weather forecasting be our introductory example, see Harper (2008).

Weather is no structure carved in stone, but exhibits a complex changing pattern in space and time. We have masses of observational data, and we believe that we have learned from the physical science of hydrodynamics the 'correct' equations. But weather forecasting requires that we be able to solve the equations, i.e. to compute future developments from current observations. This sounds easy, but turned out to be impossible through the first part of the 20th century. A new possibility arose around 1950 with the advent of modern electronic computers. The Institute of Advance Study in Princeton became the testing ground under the leadership of John von Neumann. But there were problems, the American community was divided between the practical weather forecasters and the highly mathematical theoreticians. The latter started playing with their equations in order to make actual computations feasible, but with little progress. The rescue came from the Scandinavian countries, where you had a tradition of combining observations and mathematical theory. To go forward you need to simplify the theoretical description, but in a way faithful to a structural understanding of nature itself. And with these simplifications the weather of tomorrow became computable today.

What happened in Princeton in 1950 has had an enormous influence on later developments, not only on scientific practice, but also on our understanding of mathematics and its applicability. Behind the somewhat forbidding title *A review of numerical methods for nonlinear partial differential equations*, Tadmor (2012), you will find a highly professional survey of the current art and science of mathematical modeling and large scale computing. This survey is well supplemented by the discussion and examples of modeling in an industrial context in Bohun (2014). We shall in our essays navigate around the formalism of mathematics itself, but aim toward an understanding of how the notions of structure, syntax, and algorithms give us an insight into the nature of knowledge, not only numerical weather forecasting, but also in the miraculous left hand of Leonardo da Vinci.

Structures: Everyone seems to have some general idea of what structure, shape and form are. These concepts must in some sense be derived from what we "see" in the world, but it is no trivial task to give a precise and universally valid definition of what is meant, i.e. to give an exhaustive list of necessary and sufficient defining conditions; just try to define in unambiguous terms what a *chair* is; see the discussion of the "epistemologies of the eye" in Daston and Galison (2007). We may ask the mathematicians for help, and there is a vast body of knowledge about shape and form in the natural world based on geometry and the calculus of variation, see Hildebrandt and Tromba (1996). But this only covers part of the terrain; for a broader view see the survey of *Nature's Patterns* in three volumes on shapes, flow and branches, Ball (2009), and the mathematical introduction to *Pattern Theory* in Grenander and Miller (2007).

This is, however, not the end of the story. If we turn from what we "see" to how we "reason", we meet the logician's concept of model or structure. This is, perhaps, the most general version of the structure concept. Basic is first the notion of a 'domain', which specifies a set of objects, things, or even individuals, and next a list of 'properties' and 'relations' defined over this domain. Structures are everywhere, if you want to count your money, the structure for counting has the numbers $0, 1, 2, \ldots$ as its domain, and uses 'larger than' and 'equal' as some of the defining relations. If you want to measure your house and garden, you use elementary Euclidean geometry with points, lines and surfaces as the domain and various geometrical notions as defining properties and relations, such as 'point on a line', 'line of intersection', 'parallel', and others. But there are more structures than the elementary mathematical ones.

In the book *Representation and Invariance of Scientific Structures*, Suppes (2002), you will find a systematic development of a scientific methodology based on the logician's notion of structure with examples from physics, from several of the social sciences, and from language and brain. We shall also in later essays discuss many examples, taken from model theory and conceptual spaces, mathematics and pattern theory, language and mind.

Syntax: The syntax and grammar of language are "tools" we create in order to describe and make use of structures. Structures are what we "see", e.g. the current weather situations. For some purposes what we "see" may be an end in itself, as when we enjoy the brilliant colors of a setting sun. But if we want to predict or even to compute how the weather of tomorrow will be, more will be required. As a first step we need a language or a syntax to describe what we see. But syntax is not enough, rules are needed to describe how elements of a structure interact and how a structure evolves over time. For physicists the language is mathematics and the rules, or "laws of nature", are the basic equations of Newton, Einstein and Dirac. The words "laws of nature" seem to signify that the laws of physics and also those of "pure" mathematics are objective, eternal and therefore if true, then necessarily so. They are not invented by man.

Against this view we may argue that mathematics and physics are products of human culture in exactly the same way as language, music, and rules of law are. We are much in sympathy with the great Dutch mathematician Brouwer who in his 1907 thesis characterized mathematics as a free activity of the human mind; thus mathematical truth is neither to be secured by some abstract platonic realm beyond space and time, nor by reference to a formal game of axiomatics and consistency proofs; see Chap. 6. But with Brouwer's exclusive focus on the individual mind there is a difficulty of explaining the applicability of mathematics to the world around us. Perhaps, there is a way out. Between the individual mind and a platonic realm of structures, there is culture, or—in the language of some anthropologists—there is the collective mind of the species understood as a social network of individual minds across which cognition is distributed; see the discussion in Gamble et al. (2014). Our particular concern is mathematics; in the last chapter we shall argue for a cultural approach to the foundation of mathematics. Many mathematicians but few philosophers agree to a cultural approach to mathematics; we shall return to this tension between the professionals and the philosophers in Chap. 9. In our discussion we shall insist that "abstract objects" such as numbers are indeed created in the interaction between cultural space and individual mind, explaining, as we will further argue, the "obvious" applicability of mathematics to man and nature.

Algorithms: The word "algorithm" has become very fashionable today. But what the word describes has been with us since the early days of mankind. " Algorithms are simply a set of step by step instructions, to be carried out quite mechanically, so as to achieve some desired result. … Algorithms are not confined to mathematics … The Babylonians used them for deciding points of law, Latin teachers used them to get the grammar right, and they have been used in all cultures for predicting the future, for deciding medical treatment, or for preparing food. Everybody today uses algorithms …" "This is a quote from a highly recommended book on the history of algorithms, Chabert et al. (1999)".

The subtitle of the book is "from the pebble to the microchip". The step from counting with your fingers to manipulating apps on your mobile phone is enormous. But is the step so large and of such fundamental importance that computing is going to be a new and fourth scientific domain alongside with the physical, biological and social sciences? An affirmative answer is given in a recent book *On Computing*, Rosenbloom (2012). The author gives many examples of computing innovations such as immersive display technologies, neurally controlled prosthetics, and advanced unmanned aerial vehicles. These are examples of so-called "augmented reality systems", located as they are at the interface between real and virtual environments. They are also examples of systems where computers replace humans. But is computing a new basic science, or is it just another methodology to be added to the researcher's toolbox? Systems of augmented reality are surely important steps forward, but do such examples define "computing" as a new branch of science? The basic textbook in computing at the Massachusetts Institute of Technology, Abelson et al. (1985), begs to differ:

> Underlying our approach to this subject is our conviction that 'computer science' is not a science and that its significance has little to do with computers. The computer revolution is a revolution in the way we think and the way we express what we think. The essence of this change is the emergence of what might best be called procedural epistemology – the study of the structure of knowledge from an imperative point of view, as opposed to the more declarative point of view taken by classical mathematical subjects. Mathematics provides a framework for dealing with notions of 'what is'. Computations provide a framework for dealing precisely with notions of 'how to'.
>
> Abelson et al. 1985, p. xvi.

This point of view is largely in agreement with our analysis of the mathematical sciences. Rather than a division into sub-disciplines such as mathematics proper, statistics and computing, we prefer an analysis based on structure, language and algorithms. Computations have always been part of the classical mathematical subjects, what is new is that computational feasibility has become as important as pure structural insight; see Fenstad (2002). This is a development partly driven by the advances in computer technology. The notion of algorithm, and hence of what is in principle computable, is an "absolute" and independent of any machine architecture. But the computational device is important: What is possible to compute, given limited time and resources, depends upon the machine; recall the example of weather forecasting at Princeton in the early 1950s. And there will always be new machines. Beyond the "supercomputers" of today, we see a promise of even more powerful "quantum computers".

The problem about computational feasibility—what is computable in real time—has become more urgent with the recent advances in the biological and the cognitive sciences. First an observation on recent work in biology and the life sciences: Here the new term is "bioinformatics", describing an activity which mixes biology and computing in order to extract new structures and patterns from vast amounts of data, see Stevens (2013). Next a remark on the cognitive sciences: Here the catch words are "artificial or machine intelligence" and "machine learning". It is remarkable how much of human cognitive skills that have been reduced to, or replaced by algorithms and "intelligent" machines, see the recent surveys in *Nature* (2015) and *Science* (2015). But further progress in both biology and the cognitive sciences demands efficient algorithms based on "big data" inputs. The situation reminds us of numerical weather forecasting in the 1950s. At that time both nontrivial structural simplifications and increased computational power proved necessary in order to succeed. New structural insight is best left to the creative scientist, here we restrict ourselves to a few words on computability and big data.

Computational success has largely been restricted to two cases, on the one side the continuous, or infinite case, based on differential equations and approximations, on the other side "small" combinatorial systems with few basic elements, based on counting and effective enumerations. In the middle ground where there are many units, such as neurons in the brain or positions in a game of Go, we meet a computational barrier. In some cases we can pass from a large discrete system to a continuous or infinite model and use "standard" methods, physicists and economists often do this. This strategy is not always possible in the social and cognitive sciences, see Chap. 3. Combinatorial complexity thus seemed to be an absolute barrier. Faster machines, even "quantum computers", will help, but are not the final answer to problems of "exponential growth" and computational complexity; new algorithms will also be necessary. Recently we could read in *Nature* an article on *mastering the game of Go with deep neural networks and tree search*, Silver et al. (2016). This particular analysis is a masterful blend of structures and algorithms solving what at the present time seemed to be an intractable problem. The success of this analysis should not let us forget that computing with neural networks has a long history, but what the example of Go shows is that this kind of computing now has matured and become an efficient tool in dealing with big data problems; for an accessible introduction to machine learning and "big data" the reader may consult Marsland (2015). We take note of only one point: What was previously done by the external user, such as the resetting of parameter values in the exploration of possible developments of the system under study, has now become internal to the computation and can be executed with machine speed. This allows us to tackle bid data problems, but we should not forget that what we obtain is critically dependent upon the "learning rules" written into the program. This is a development that, however, fits well in with our general analysis, and we may safely predict that machine learning and similar techniques will be an essential ingredient of the toolkit of future practitioners of the art and science of mathematical modeling.

Algorithms and computations are of rapidly increasing importance in science and society. The "wild new world of data", to quote a reviewer in a recent issue of *Nature*, has as seen above an obvious impact on science and technology, but equally important are the social implications. A thoughtful discussion of the social aspect is found in a recent book *Data and Goliath: The Hidden Battles to Collect Your Data and Control Your World*, Schneier (2015).

Most readers will have had some exposure to computers, rules and algorithms. Indeed, every owner of a current smartphone is an expert user of such tools. But in addition to practical skills there is a need for a deeper understanding of the "computer" revolution. This has been our concern so far in this chapter. For additional reading we recommend for the general reader the book *Algorithmics—The Spirit of Computing*, Harel (2012). If you want to see some of the mathematics involved, you may consult the survey *Computability Theory: Structure or Algorithms*, Fenstad (2002).

Mathematics is not only a body of theory, it is also a human activity. We conclude this part on structure and algorithms by a brief remark on the various roles played by the mathematicians in plying their craft. We may speak of "the hired hand" versus "the toolmaker", a division we prefer to the standard one of applied versus pure mathematics. This division is not absolute, most mathematicians have bit parts in both roles. It is not to be denied that many, perhaps, even most "pure" mathematician see their subject as an autonomous field of knowledge and would resist our view that mathematics is but the link between structure and algorithms. However, in order to appreciate the necessity of mathematics in an analysis of the nature of knowledge, we need to understand both the toolmaker and the hired hand and how they interact in grasping the complexities of structure, syntax, and algorithms. To explain this is a major challenge, which, we believe, is not well communicated today to common man. Mathematics is too often seen as an esoteric activity, a game for the selected few and irrelevant for the many. We will in these essays argue for the necessity of a broader perspective on mathematics and knowledge.

1.3 Mathematics and Knowledge

Is mathematics, as characterized in the section above, necessary for a general understanding of knowledge? We have above argued for the importance of the concepts of structure, syntax and algorithms. In doing so we have, in fact, presented a geometric version of standard HD (hypothetic-deductive) methodology. We have started with what we "see", i.e. with structures, and regarded syntax and formalisms as tools to describe and control structures. Standard expositions of logic and scientific methodology turn this around, recall our discussion above of the so-called "experimental logic" of Wilczek: First we fix syntax and rules of deduction, next we add a collection of hypotheses to the axioms and rules of logic and mathematics. Then we start the game of "playing with the equations", seeing what consequences there may be. The ideal outcome should be a "law of nature", but since such laws usually have a universal form ("all swans are white"), there are problems with verification and truth. Testing scientific theories is a large sub-discipline of the philosophy of science, see e.g. Earman (1983). We shall not deny that the syntactical approach to methodology has its advantages. It is a precise theory and has an impressive mathematical superstructure. But is "playing with the equations" the way science works?

In the geometric approach we have used mathematical modeling as our paradigm. Starting with structures is not, as noted above, a novel point of view, see Suppes (2002) as well as the discussion in Chap. 3. I have also noted with interest the cognitive approach developed by Giere (1988). The use of structure in the present account corresponds to a large degree to his use of models, and we both agree that knowledge is grounded in both brain and culture. We recommend his text for a number of issues not sufficiently discussed in this chapter, such as the relationship between abstract models and the "real" world, or in other words, the link between what we see and what there is. The praxis of mathematical modeling will be spelled out in more details and illustrated by many examples in the following chapters; let us at this point just give a quick preview: Modeling consists typically of three stages: First an analysis of a given scientific or technological problem leading up to a well-defined structure (notice that we in this sentence have hidden all what we have to say about observations and experiments); next comes syntax and "equations", which are used to analyze the properties of the structure; finally the development of algorithms, which in the scientific context can be used to make specific predictions and to determine long-time developments, and which in a technical context is used to produce "blueprints" for construction and production. But equations normally have their limitations, they cannot always be exactly solved. This adds a further ingredient to the practice of mathematical modeling, simulations and graphical displays.

Such is the simple picture. In real-life modeling there is usually a fair amount of interaction between the three stages of structure, syntax and computations. The environmental sciences present a good example of this complexity, see our initial example of numerical weather forecasting at Princeton in the 1950s; a current review is found in Wainwright and Mulligan (2004). We should also take note of the complex relationship between what is immediately measured and what is reported as data in many experimental situations, PET (positron emission tomography) scanning as a tool of cognitive sciences is a good example, see the discussions in Chap. 3 and also in Humphreys (2014).

We should also at this point make a few comments on the duality between the structural/semantic and the syntactic approach to general methodology. The study of this duality is a key point in the work of P. Suppes, see Suppes (2002). For a logician the prime example is the Gödel completeness theorem for first order logic, where validity/semantics is equated with provability/syntax. For Suppes it is the theory of measurement that stands in focus, and where representation theorems and invariance arguments play a central role in connecting qualitative and numerical measurements; see the analysis in Friedman (2014). I have used a similar approach in the analysis of the concept of probability in Fenstad (1968), where a representation theorem provides a link between Carnap's two concepts of probability by showing that logical probabilities, being the linguistic approach, always can be understood in terms of urn models and observed frequencies, being the structural approach. Adding suitable symmetry conditions the link is precisely the celebrated B. de Finetti's theorem on exchangeable events. Representation theorems are always welcome, but they do not always exist. In systems extending first order logic there are always non-intended models. This is a consequence of the Gödel incompleteness theorem and proves that the model-theoretic or semantic approach cannot be reduced to syntax—axioms can never completely characterize what you see.

In the following chapters there are several illustrations of the geometric approach taken from the physical, biological and social sciences. We shall also in Chap. 3 add some general remarks on how structures evolve over time (classical/linear, stochastic, non-linear). What we "see" is not the full story of how science works: there will be some additional remarks in Chap. 4 on reductionism and causality in complex systems. We should also add a word of warning to the reader: As pointed out above our use of the word "structure" is consistent with its use in mathematical modeling. This usage is related to, but does not correspond exactly to its use in what is known as "structuralism" in the philosophy of mathematics; for a careful discussion of the latter, see Burgess (2015).

But is the structure of knowledge as seen in the physical, biological and social sciences a model for all kinds of knowledge, even of knowledge and insight in the human domain? This is a difficult and highly controversial issue. But one point is clear. Structures are prior to algorithms, and insight and creativity is needed to give a "correct" or productive structure to what is "seen". And given a structure, not every question asked can be effectively answered by an algorithm. This follows from general results on computability and undecidability proved by Gödel and Turing in the 1930s, see Harel (2012); for a recent example from physics see Cubitt et al., *Undecidability of the spectral gap* (2015). Thus an approach based on the mathematical way, combining structures and algorithms as we see in the current art and science of mathematical modeling, does not necessarily make man into mechanical man.

We further note that the word "knowledge" can point to many things, from the syntactic and proof-theoretic study of some operator K in certain systems of formal logic to the cognitive scientist's study of mind-states characterized by such words as "insight" and "understanding". I would like to believe that the "mathematical way" toward knowledge as presented above is valid almost everywhere, including art and the humanities. We need to understand this claim in a proper way. In very few cases are structure, syntax and rules so simple that the future is exactly computable, one example is space travel. Earthquakes are different. We may have detailed structural insight and vast computational power, but the exact time and location of the next quake is not computable. In such cases we may use structure and algorithms not to predict, but to give advise on how to prevent adverse outcomes. In the social sciences and the humanities we may have much computational power but not sufficiently precise structural insight in order to predict or even to give advice. But this does not in any way deny the importance of structure, syntax and algorithms in our effort to understand what knowledge is. We shall present a number of examples to support this claim.

We first offer an example from the history of art and science, see Chap. 2 on the miraculous left hand of Leonardo. We should note that in historical studies both in art and the sciences the main effort is often directed towards establishing the "correct" original structure and how it developed. A typical case bridging the humanities and the sciences is the study of how language evolved, see the survey in Fitch (2010). Many other examples can be found in anthropology and the geological sciences. The challenge here is to fit many single bits of information into a coherent and total structural view; there is almost never a unique solution to this problem. Workers in these fields persist, but philosophers of science always argue at length whether such activity is science or just storytelling.

From Leonardo we proceed to the cognitive sciences, in particular to language and meaning. We tell a story with three parts: first, on the origin of syntax and how this story depended on what we "see"; second, on the many interactions between grammar and logic, with some remarks on current natural-language technology, see Chaps. 5 and 7; and third, on formal semantics, geometry and mind. Grammar, logic and mind are covered in Chap. 8; see also the discussion in Gärdenfors (2014). Some remarks on structures and how proto-syntax emerged is found in Chap. 9; for a more extended analysis of how language developed, see Fitch (2010) and also the book *Grammar, Geometry, and Brain*, Fenstad (2010).

Through concepts such as structure, language, and algorithms I have tried to argue for the importance of the "mathematical way" in today's knowledge society. We should remember that this is really an ancient art. From simple applications to the stars in heaven and the cathedrals on earth we have arrived at the current art and science of mathematical modeling, applicable to a vast range of phenomena "seen". This also includes many cognitive tasks. It is, however, impossible to know what exactly can be reduced to algorithms and intelligent systems. But thanks to Gödel and Turing, and granted enough human wisdom, the tools will never become the masters.

References

Abelson, H., Sussman, G. J., & Sussman, J. (1985). *Structure and interpretation of computer programs*. Cambridge, Massachusetts: The MIT Press.

Ball, P. (2009). *Nature's patterns: Shapes, flow, branches* (Vol. 3). Oxford: Oxford University Press.

Barrow, J. D. (2010). Simple really: From simplicity to complexity—and back again. In B. Bryson (Ed.), *Seeing further, the story of science and the royal society*. London: Harper Press.

Bohun, C. S. (2014). An Introduction to modern industrial mathematics. *Notices of the American Mathematical Society, 61*, 364–375.

Burgess, J. P. (2015). *Rigor and structure*. Oxford: Oxford University Press.

Chabert, J. L., et al. (1999). *A history of algorithms, from the pebble to the microchip*. Berlin, Heidelberg, New York: Springer.

Cubitt, T. S., Perez-Garcia, D., & Wolf, M. M. (2015). Undecidability of the spectral gap. *Nature, 528*, 207–211.

Daston, L., & Galison, P. (2007). *Objectivity*. New York: Zone Books.

Earman, J. (Ed.). (1983). *Testing scientific theories*. Minneapolis: University of Minnesota Press.

Fenstad, J. E. (1968). The structure of logical probabilities. *Synthese, 18*, 1–23.

Fenstad, J. E. (2002). Computability theory: Structure or algorithms. In W. Sieg et al. (Eds.), *ASL Lecture Notes in Logic: Reflections on the foundation of mathematics*. Natick, Massachusetts: A K Peters Ltd.

Fenstad, J. E. (2010). *CSLI Lecture Notes 200: Grammar, geometry, and brain*. Chicago: University of Chicago Press.

Fitch, W. T. (2010). *The evolution of language*. Cambridge: Cambridge University Press.

Friedman, M. (2014). An overview: Suppes as a scientific philosopher. In C. E. Crangle, A. Garcia, & H. Longino (Eds.), *Foundations and methods from mathematics to neuroscience*. Stanford: CSLI Publications.

Gärdenfors, P. (2014). *The geometry of meaning*. Cambridge Mass: The MIT Press.

Gamble, C., Gowlett, J., & Dunbar, R. (2014). *Thinking big: How the evolution of social life shaped the human mind*. London: Thames and Hudson.

Giere, R. N. (1988). *Explaining science, a cognitive approach*. Chicago: University of Chicago Press.

Grenander, U., & Miller, M. (2007). *Pattern theory: From representation to inference*. Oxford: Oxford University Press.

Harel, D. (2012). *Algorithmics: The spirit of computing* (3rd ed.). Berlin, Heidelberg, New York: Springer.

Harper, K. C. (2008). *Weather by the numbers: The genesis of modern meteorology*. Cambridge Mass: The MIT Press.

Hildebrandt, S., & Tromba, A. (1996). *The parsimonious universe: Shapes and form in the natural world*. New York: Springer.

Humphreys, P. (2014). Models of data and inverse methods. In C. E. Crangle, A. Garcia, & H. Longino (Eds.), *Foundations and methods from mathematics to neuroscience*. Stanford: CSLI Publications.

Machine intelligence. *Nature, 521*(7553), 435–458, 28 May 2015.

Marsland, S. (2015). *Machine learning, an algorithmic perspective*. Boca Raton, Florida: CRC Press, Taylor & Francis Group.

Rise of the machines. *Science, 349*(6245), 260–278, 17 July 2015.

Rosenbloom, P. (2012). *On computing—The fourth great scientific domain*. Cambridge Mass: The MIT Press.

Schneier, B. (2015). *Data and Goliath: The hidden battles to collect your data and control your life*. New York: W.W.Norton.

Silver, D., et al. (2016). Mastering the game of Go with deep neural networks and tree search. *Nature, 529*, 484–489.

Suppes, P. (2002). *Representation and invariance of scientific structures*. Stanford CA: CSLI Publications.

Stevens, H. (2013). *Life out of sequences: A data-driven history of bioinformatics*. Chicago: University of Chicago Press.

Tadmor, E. (2012). A review of numerical methods for nonlinear partial differential equations. *Bulletin of the American Mathematical Society, 49*, 507–554.

Wainwright, J., & Mulligan, M. (Eds.). (2004). *Environmental modeling: Finding simplicity in complexity*. London: Wiley.

Wilczek, F. (2002). A piece of magic: The Dirac equation. In G. Farmelo (Ed.), *It must be beautiful: Great equations of modern science*. London: Granata Books.

Part I
Structures and Algorithms

Chapter 2
The Miraculous Left Hand—On Leonardo Da Vinci and the Search for a Common Understanding of Man and Nature

Abstract Is a common approach to knowledge about man and nature possible? With Leonardo da Vinci as our starting point we will explore this question. Leonardo was much more than a painter; he was a sharp observer of man and nature. What he saw (structures) and what he did (using his miraculous left hand as syntax) was translated into insights and methods of construction as recorded in his notes books. His view was holistic, as witnessed by his study of the eye; from the anatomy of the eye he proceeded outwards to a theory of perspective, and inwards to an understanding how perception and mind are grounded in the brain. In this we see similarities to current studies of language, mind and brain. It is possible to see a kind of proto-version of the art of mathematical modeling in Leonardo. This is a methodology of wide scope, extending far beyond physics and engineering. With a sufficiently broad understanding of the key concepts of structure, syntax and algorithms we have a method strong enough to allow for a common approach to knowledge of man and nature, to what there is and how we know.

Keywords Common understanding of man and nature · Anatomy of the eye
Leonardo—mathematical modeling

Is a common understanding of man and nature possible? Science, in the broader sense of "Wissenschaft", seems to say no, negating any attempt by man to achieve a unified view of knowledge. What we observe is a science progressing in many, perhaps, even contradictory directions. Many would insist that this is as it should be. Science is a growing body which cannot be constrained by any a priory scheme of thought; the unity of science is and should be problematic. From any local point of view this may be true. But in a broader historical and cultural perspective we see traditions where different disciplines have interacted to create a common environment of learning and practice. And it is the experience of such cultural traditions which has justified a belief in the unity of knowledge.

We may point to one dominating tradition in our Western culture where mathematics, philosophy and our understanding of nature came together in a broad stream flowing from Antiquity to Kant and Laplace. This tradition had an enormous influence; we cannot, to mention one important example, understand the cartesian methodology, its range and limitations, without at the same time having an insight into Descarte's

J. E. Fenstad, *Structures and Algorithms*, Logic, Argumentation & Reasoning 15,
https://doi.org/10.1007/978-3-319-72974-9_2

analytic geometry and mechanistic physics. And the mathematical framework of this development had an important impact on the general theory of knowledge. It informed, but it also set limits, not always understood, on our understanding of key concepts such as determinism and causality as seen in processes both in nature and society.

Another tradition is represented by the growing understanding of man as a biological species, covering the development from Darwin to current population genetics and molecular biology. This is a second example where we see many disciplines and traditions coming together, and it is only through a sound insight into this historical development that we can judge both the strength and the limitations of this process. Too many people are mislead by the grand words and splendid rhetorics of vulgar darwinism and become easy preys to the social darwinism of earlier times and the sociobiology of today.

However, an understanding of the unity of knowledge requires more than physics and biology. It must also "explain" man as a thinking individual, as a carrier of meaning and consciousness.

The cognitive side was not entirely absent from the earlier history. Let us take grammar as our example. A knowledge of grammar and logic was an important part in the first stages of the development from Antiquity to Kant. But this did not last. We observe that the *Grammatica Speculative*[1] of Thomas of Erfurt from 1315, a strong brew of language, logic and philosophy, has disappeared from our shared historical consciousness, as later the grammar of the Jansenists of Port Royal, the latter known as the refuge of Pascal. We remember only a "royal list" of natural scientists, Euclid, Ptolemy, Copernicus, Kepler, Galileo, and Newton. But the situation is changing. The science of language underwent a major development starting with the dane Rasmus Rask and the comparative linguistics of the mid 19th century,[2] continuing with de Saussure and structuralism from the beginning of the 20th century, then the theoretical linguist Chomsky in the middle part of the 20th century, to the developments in language and speech technology that we see today.[3] For many people language is the unique characteristics of man as a biological species; thus if the current science and technology is successful, this should offer some promise of including the study of man and meaning in the broader science of man as part of nature. Let only the molecular biologists and the speech technologists get on with their work.

Not everyone would agree. In the same way as the chess-playing machine is not a human being, a speech-generating machine is not an individual with consciousness and the capacity for meaningful language. Thus for many the initial question-mark will stand. Mind and body belong to different categories, for how could my consciousness and my understanding of myself be captured by any combination of measurements of time, length, and mass. The humanities and the natural sciences,

---

[1]Thomas' book is reprinted together with an English translation in Bursill-Hall (1972). For a broader analysis of medieval grammar see Paley (1976).

[2]Pedersen (1972) is our source for Rasmus Rask and linguistics in the 19th century.

[3]Chomsky (1957). I have given my view of the development of linguistics from Thomas of Erfurt to Chomsky and the language technology of today in Fenstad (2010).

both of equal importance, must in their methods go their separate ways. The dream of a common way of understanding has, however, deep roots in our culture. Much has been written and much can still be said about this, and the literature is enormous. But I note, with some surprise, that there is one man who stood right at the beginning of the quest for a common understanding, and who is often not given the proper recognition, Leonardo da Vinci.

LEONARDO DA VINCI. To the general public Leonardo da Vinci is primarily known for the portrait of Mona Lisa, a high point in Western art. Less known is the scientist Leonardo and his view on the nature and unity of knowledge.[4]

Leonardo's formative years were spent in the workshops and studios in Firenze as an apprentice in many trades. He was left-handed, not uncommon among men of talents, and his "miraculous left hand" made it an easy task for him to reproduce in drawings and paintings what he was working on, and more generally, what he observed. In the earlier stage, as an apprentice, we see mere reproductions in his notebooks. But with experience and practice he moved beyond, and many new inventions of his own started to make their appearance. He did not have a mathematical way of expressing his thoughts and observations, as did the astronomers of Antiquity and the scientists of the later Renaissance period; he observed and he reproduced what he saw. Water and systems of water supply were an important domain of activity for the young technologist Leonardo, and in the notes he left we find several sharp observations of hydrological processes. But despite the many achievements of Archimedes, the time was not ripe for a mathematical analysis of "real" hydrological phenomena. Later Leonardo's interests expanded to include natural processes in geology and the living nature, the latter also including man's body and soul. Gradually he developed a vision how to understand the interconnectedness of things. He wanted in the same framework to obtain an insight into the mechanical principles behind the many inventions and tools of the engineer, and into the corresponding "mechanical" principles governing processes in nature and man. His studies of how water flowed inspired him to search for a more general theory of how objects and forms are transformed. Again he lacked a mathematics strong enough to describe what he "saw", but in several series of sketches from his notebooks we can, with generous hindsight, see the beginning of a geometrical transformation theory.[5]

To "see" was the method of Leonardo, and in this he made a break with the traditional learned style of the Renaissance, where interpretations of the newly discovered text from the classical Greek period was the chosen method; we recall many illustrations in the medical texts from the anatomy theaters of the Renaissance where the learned professor reads from an ancient text, perhaps from a recovered manuscript of Galen, when at the same time the assistant performs a dissection on a body on a table in front of the professor. For the learned man the link between text and reality, in this case a human cadaver, was rather weak, what was "seen" and what was "read" was not necessarily connected. Leonardo represented something new and different. Knowledge for him was alway grounded in what was "seen". This was a break with

[4]The literature on Leonardo is extensive; an easily accessible biography is Nicholl (2005).

[5]See the chapter "Geometry done with motion" in Capra (2007).

a long lasting tradition. It is of interest to note that even Galileo, a century later, was in many ways a follower of the old tradition, where an "experiment" more often was a literary illustration of an accepted truth than something immediately "seen".[6] Returning to Leonardo, we note that the anatomy of the eye became a high point in his studies of man. He followed his anatomical investigations of the eye outwards, from the primary fact of seeing to a search for a structure for what was seen. And in this process he created a new theory of perspective to serve as a foundation for his art and more generally for his understanding of nature. He also pursued his studies of the eye inwards into the brain to search for an anatomical foundation for man's emotions and cognitive abilities. And the "outwards" and the "inwards" should be connected. Leonardo as scientist moved far beyond the painter described by current traditions. And different from the later Descartes, Leonardo was no dualist making a sharp distinction between body and mind.[7] His work can be read as an argument for the unity of the two, with a mind firmly grounded in the anatomy of the brain.

Later times have not always seen Leonardo in this perspective. It is true that some enthusiastic supporters of Leonardo the scientist, such as the French philosopher of science and historian Pierre Duhem, have given Leonardo a prominent role in the development of modern science. More sober observers, such as the well known historian of science E. J. Dijksterhuis, remained unimpressed. In his major study *Die Mechanisierung des Weltbildes* he very briefly notes that Leonardo was of almost no importance compared to the great names of the scientific revolution.[8] Today we allow for a somewhat more generous view. Leonardo's originality as engineer is convincingly documented in P. Galluzzi's rich book *Renaissance Engineers from Bruelleschi to Leonardo da Vinci* from 1996. Galluzzi, who is Director for the Istituto e Museo di Storia della Scienza in Firenze, write with rare insight and authority about Leonardo. Leonardo's science has also been the subject of two recently published books *The Science of Leonardo* by Capra 2007 and *Leonardo's Legacy* by Klein 2010. Both applaud, perhaps a bit too enthusiastically, the scientist and his attempts to understand complex systems, and both deplore the lacking recognition of Leonardo as a pioneer scientist. But the latter neglect may have a simple explanation. Later times knew Leonardo only as painter, his notebooks and drawings were for a long time not published. Thus it is A. Vesalius who has been recognized in the standard histories of medicine as the creator of modern anatomy, whereas the pioneering work of Leonardo of equal importance was hidden in his forgotten notebooks.[9]

Leonardo was much more than a painter and illustrator. He therefore merits an honorable mention in the history of science supplementing the "royal list" of the scientific revolution. But we must be aware of his limitations. We may rightfully

[6]Here I touch upon a topic which is extensively discussed in Dear (1995).

[7]For an analysis of the dualism of Descartes from the perspective of current cognitive science see Clarke (2003).

[8]The sources are Duhem (1906–1913) and Dijksterhuis (1956). I recommend Turner (1995) for a critical review of the changing conceptions of Leonardo and his work.

[9]Debus (1978). Debus comments in length on Vesalius, Leonardo gets just one sentence: Leonardo da Vinci's (1452–1519) earlier anatomical drawings were masterful, but unfortunately of little impact as they were unpublished.

recognize that his morphology for complex structure was remarkable at the time, but today his notebooks and drawings are mostly of historical interest, and not a working source for current science. This is, however, not the end of the story. If the artist Leonardo and the scientist Leonardo both had been part of a living memory today, there would have been no need for C. P. Snow to write his book (Snow 1959) on the two cultures.

WHAT DID LEONARDO "SEE"? To meet for the first time the sketches and paintings of Leonardo is a powerful experience. It is far beyond my competence to describe and analyze his art in any details; the reader would be well advised to consult the sources I have mentioned in the section above. But a mathematician may have something different and new to add to the standard expositions of what Leonardo "saw".

Mathematics often appears to the common man as a vast collection of numbers, equations and complex geometrical constructions, to be learnt on command and to be applied with more or less skill in everyday tasks of counting and measurements. But mathematics is not what the common man believes, as I will try to explain. A key concept necessary for understanding mathematics, is the notion of *structure*.[10] To be brief, a structure consists of two parts. One is the *domain* of the structure, which is some given collection of things, objects, or individuals of any kind, be they of material, mental or cultural origin. The second part consists partly of *properties* that the objects are supposed to possess, and partly of *relations* these objects bear to each other. "Common man" will correctly suspect that this is how a mathematician writes, and it is true that the simplest examples of structures come from mathematics, but the concept has, as later parts of this essay will show, a much vaster range, not only including mathematics and the natural sciences, but also art and the humanities.

One simple example of a structure is the collection of natural numbers 0, 1, 2, 3, …, where the defining relations are addition and multiplication; some basic defined properties are even and odd and being a prime number. Essential for all kinds of measurements and construction are geometrical structures based on points, lines and surfaces. And for such structures length and distance are fundamental properties and relations. School mathematics knows its geometry through the heritage from Euclid's Elements with its characteristic axiomatic approach. But geometry was not only an intellectual activity of the learned class, the artisans and the master builders knew their geometry as part of their trade, as the cathedrals of the medieval ages bear witness to.

It was to this latter tradition Leonardo belonged. Indeed, it was in his early years as apprentice in the workshops in Firenze that he got a "grasp" of what structures in essence are. We are impressed by his sketches and notes on running water and on many kinds of mechanisms connected to waterworks. We are equally impressed by

[10]The most general analysis of the notion of structure is today found in mathematical logic. This analysis has an important root in the work of the Norwegian logician Thoralf Skolem, see Fenstad and Wang (2009). But the logic approach may at times be somewhat too restrictive. An alternative would be to see the notion of structure in light of prototype theory, which in a sense is to understand structure as a kind of "fixed point" for what is seen; see Rosch (1978) and Gärdenfors (2000), see also the discussion in Fenstad (2010), pp. 45–46.

his studies of geological forms and structures; sometimes in his paintings he seems to pay even more attention to the landscape and the mountains in the background than to the people in front. And we note that the structures represent Leonardo's own interpretation of what he "sees", when he for example does not give the "correct" reproduction of light and shadows, but uses the shadows to open up for a deeper insight into the character of the person he is painting. The interested reader should turn to the discussion in Klein (2010) of Raphael's *Portrait of a Lady with a Unicorn* compared to Leonardo's *Mona Lisa*.

To give what is "seen" a structure was for Leonardo a first step towards insight and eventual applications. He was a part of an artisan tradition which lived side by side with an emerging mathematical natural science and technology. For the role this artisan tradition played within engineering I recommend the book *Engineering and the Mind's Eye* by Ferguson (1993). He shows through a series of examples how the engineering practice traditionally, and even today, went well beyond the purely scientific calculations possible at the time and included qualities such as how structures are "seen" and how qualities of materials are judged, i.e. properties solidly within the domain of the Mind's Eye. I would guess that boat designers of today would still agree to this, even if computers are fast approaching. For an example from the history of natural science I refer to the study *Image and Reality* by Rocke (2010), where it is convincingly documented how a visualization of chemical structures in the mid 19th Century, above all seen in the work of the German chemist A. Kekulé, started a structural revolution going well beyond what "official" theory of that time could sanction.

THE MIRACULOUS LEFT HAND. The structures seen by Leonardo, were not only structures frozen in time. Many examples from his notebooks show structures undergoing a time development, in particular, the many examples he drew from hydrodynamics and geology. But the act of "seeing" structures is only the first step on the road to a complete understanding and to a knowledge necessary to make use of them. To be able to manipulate and control structures we need an appropriate language or syntax. It was exactly on this point that the "royal list" of Kepler, Galileo and Newton, made a decisive break-through in our understanding of nature. Their glory was to create a mathematical description of how bodies move, valid both for planets in the sky and cannon-balls on earth. To appreciate this it is important to observe a separation between syntax and structure. Galileo's equations for falling bodies, Kepler's equations for the orbit of planets, and the laws of force written down by Newton are linguistic expressions for what is "seen" in nature. And this mathematical formalism also included a description of the long term time development of the structures, thus allowing us to predict when a spacecraft will land on the moon, or how to aim our cannons against a foreign fortress. It is important to stress that syntax presupposes structure. The equations we write down, are always an expression of some properties of the structures "seen". Thus if structure is changed, as for example when we move from classical to relativistic mechanics, syntax including rules and equations must also be rewritten.

The syntax Leonardo used was in a special way embedded in his "miraculous left hand". He could not, as Galileo did, write the equations for a ball rolling down an inclined plane. But he could reproduce through connected series of sketches both the static and dynamic properties of what he observed. And true to his background as craftsman and engineer it was the everyday and complex phenomena that caught his attention. Leonardo had a strong interest in mathematics, but mathematician he never became despite many tutorials from close mathematical friends. We must also, in order to be fair towards him, admit that a mathematical syntax for the kind of phenomena that interested him, was of such a complexity compared to the pendulum and the inclined planes of Galileo, that we only today can see the beginning of a satisfactory analysis. But the eye of Leonardo was so sharp that "he grasped connections using nothing more than his pencil that it took other scientists many generations to formulate in equations".[11] This observation (see Klein 2010, p. 181) by a well respected professor of bioengineering at Caltech, necessitates some further remarks on mathematics, or rather on the combined art and science of mathematical modeling. We must then add a new concept to the familiar pair of structure/syntax; this is the notion of algorithm. The act of modeling has three stages: First, an analysis of a given scientific or technological problem; through this analysis we are led to a suitable structure of the problem. Next, the determination of an appropriate syntax, including equations or more generally rules, in order to describe and analyze the structure. Finally, the development of algorithms, today mostly given in the form of (computer) programs, which in a scientific context can be used to predict later stages of the evolution of the structure, and which in technological applications can be used to provide blueprints for construction and production.

But equations have their limitations, we have learned that it is only in very special cases that they are exactly solvable. In all other cases we have to resort to simulations and graphical representations of the solutions. This represents in many ways a "third way" to understand nature, where the first two are theory and experiments. Climate and environment offer good examples. We do have, at least in principle, the correct equations from hydrodynamics. Starting from these equations Vilhelm Bjerknes wrote down in the first years of the 19th century a set of equations for weather prediction. Bjerknes was guided by a deep insight into the structure and syntax of the phenomenon he studied. But his equations were complex (necessarily of non-linear nature) and could only be applied through simulations and graphical representations.[12] And it is at this point that we see a link back to Leonardo. He had a good command of two of the parts of mathematical modeling, structure and graphical representations. He lacked a mathematical syntax, hence the somewhat

[11] A current example of this can be found in Eloy (2011). Eloy writes down the necessary equations to explain Leonardo's observation that the total cross-section of a tree remains the same along its height every time it branches.

[12] This is for our purposes an important example. Bjerknes and the Bergen school in meteorology with its root both in the practice of actual weather forecasting and in the physics of hydrodynamical phenomena, got the two first parts correct, i.e. the structures and the equations, but it was only with the advent of modern computing around 1950 that the last part came into existence and algorithms were developed and implemented to enable real-time weather prediction. Numerical weather forecasting was an early example of mathematical modeling based on structure, syntax and algorithms; see Harper (2008) for a full account.

negative remark from Dijksterhuis[13] on the limited importance of Leonardo for the development of the mechanical sciences. But what Leonardo did points, as we shall in the next section, towards a broader applicability.

Reviewing the three basic concepts *structure*, *syntax* and *algorithms*, we come to understand how the general notion of structure, as described above, and the extended notion of algorithm or program, well-known to everyone who ever used a computer, also extend our possibilities for choice of syntax. It is therefore possible today to proceed with this mode of analysis far beyond the equations of physics, perhaps even to such core areas of the humanities as language and meaning. We shall in the last section reflect on what this implies for our search for a common understanding of man and nature.

A COMMON UNDERSTANDING. Leonardo was no dualist. Based on his analysis of the anatomy of the eye and how it functioned, the soul was for him in a direct and concrete way located in the brain. He observed, he made sketches and notes, but despite his deep understanding of the complexity of nature, this was but suggestions and beliefs. The search for a common understanding was still a distant goal.

Science progresses through doing what is doable. The winning strategy of the "royal list" was to simplify, to abstract from the full complexity of nature and to single out some specific aspects that could be studied in depth. Thus Galileo restricted himself to observe running balls on an inclined plane where Leonardo aimed at a full understanding of the behavior of "real" water in actual rivers and waterworks. The simplifying strategy won the day, and modern natural science grew out of the simple equations written down by Galileo, Kepler and Newton. The main ingredients became experiments, analysis and synthesis. And the strategy for the understanding of more complex situations would first be to break down a complex problem into its simplest component parts and to study these parts in isolation, then as the next step to determine how the overall behavior was a well defined function of the operations of the component parts. It is not surprising that the success of this approach led to a wide-spread belief in a general reduction of man and nature to nature basic parts, i.e. to physics and physics alone. As late as 1980 we find a robust and almost unbelievable expression of this view in the Inaugural Lecture[14] delivered Stephen Hawking when accepting the professorship once held by Newton. And true to this heritage he started out by affirming his strong belief in a final and complete theory of physics. This was soon to be expected; then in orderly and neat succession would follow first a reduction of chemistry, next of biology, and finally of the study of human behavior to the basic equations of physics. Hawking had to admit that the program was not yet completed, he did with some regret note that "although in principle we know the equations that govern all of biology, we have not been able to reduce the study of human behavior to a branch of applied mathematics". Thirty years have passed since Hawking gave this lecture, and I doubt if he would have said exactly the same today. Both science and technology have progressed beyond the limits enforced by the traditional use of

[13]See Dijksterhuis (1956). Note that Dijksterhuis judges Leonardo totally in terms of classical mechanics, non-linearity and complexity had not yet reached the science historians around 1950.

[14]Hawking (1980).

linear equations; we now see the beginnings of a richer mathematics necessary for a deeper understanding of non-linearity and causal interactions in complex systems.

Let us one more time return to Leonardo. His struggle to understand nature lost compared to the strategy of Galileo, Kepler and Newton. But those who won had a prize to pay, their world became limited and colorless. And they were committed to a reductionist strategy, as seen in the quote from Hawking above. But despite this limitations we have to admit that the reductionist strategy seen in a historical perspective was a great success and became a fundamental driving force in the growth of Western society. However, in the attempt to understand the interaction between body and mind problems arose, and it is not surprising that some form of dualism became a preferred way out. If you cannot construct the soul and human consciousness from what exists at the atomic level and the forces active there, the soul must have another mode of existence. This was what Leonardo denied. His view was holistic and not reductionist. The structures he "saw" was meant to represent the phenomena in their full complexity, both "outwards" in nature and "inwards" in the brain. I do not in any sense argue that Leonardo's way should have replaced the methodology of Galileo, but I do suggest that if we had let both live and interact, we would have inherited a richer view on the relationship between man and nature; C. P. Snow's book on the two cultures, as noted above, would have been unnecessary.

The interaction between language, meaning and brain will be our counterexample to criticism of Snow. The challenge is to understand how language and meaning can be grounded in the dynamical activity of large and complex systems of neurons in the brain. The study of language must necessarily proceed on many levels. At the "top level" we need a theory of grammar including both syntax and semantics, at the "bottom level" we need a detailed understanding of the structure and the functioning of the brain. And "top" and "bottom" must be linked. To achieve this we need more than the notes and sketches of Leonardo.

And much has indeed happened in linguistics and the connected cognitive sciences.[15] New and powerful experimental and theoretical methods have been developed to help understand both linguistic and mental phenomena. We have, in particular, seen an increased use of mathematics and the natural sciences in this endeavor; mathematical modeling in the sense described above has become a standard tool. But despite this tremendous step forward we are still in a somewhat uncertain state in our understanding of the gap between language and brain. Many ways of closing this gap have been proposed. There is a clear connection between our sensual perceptions, our ability to see, hear, feel and touch, and the dynamical processes in our brain. There are also connections between grammatical categories and human perceptions. If these insights are added together with the necessary use of some mathematics, it is possible to suggest a link between language and brain.[16] This is just one example of how current science tries to ground cognitive functions in the brain. But one should show proper caution. In this area there is no lack of great visions and speculations on

[15]For a survey see Jackendoff (2002), Brown and Hagoort (1999) and Fenstad (2010).

[16]See Chap. 8.

how mind and body are linked together in one grand holistic system, but in almost all such cases this is exactly what it is, speculations and not "hard" science.

This is not necessarily the end of the story. Despite our skeptical attitude there seem to be sufficient grounds to claim that language and meaning are grounded in the brain. It is important here to have a clear idea of what we mean by the term "ground". We have gradually come to understand the notion of causality in complex systems and the limits for classical reductionism in non-linear systems. In such systems a complete theory necessitates explanations on many levels and an understanding of causal interaction between levels;[17] contra Hawking there is more than a single "upward" causal chain from physics to mind and culture. The interaction of levels does not preclude that levels may have a large degree of autonomy, and that a satisfactory explanation of a problem or phenomenon can be given within a single level. Let our example be the linguist analysis of metaphors; hardly anyone would insist that the equations needed to understand the dynamics of the brain should be present in every linguistic analysis of metaphors. Our understanding of man, meaning and nature will therefore not be a simpleminded reductionistic explanation, but could also be an understanding of phenomena at their own appropriate level; the student of literature need not start every analysis with a section on brain structure. But in the same way as all properties of thermodynamics is grounded in the molecular level and what there exists, every analysis of metaphors and meaning is necessarily linked to the human brain; the skeptical reader is well advised to turn to the discussion in P. Gärdenfors' book, *Conceptual Spaces; The Geometry of Thought*.

Our discussion of Leonardo pointed in two directions, to what he saw and to what he did. As to the first dimension, which is the level of structures, I would argue that we clearly see similarities between our discussion of language, meaning and brain and Leonardo's studies, both "outwards" and "inwards", of the eye. The second dimension, represented by syntax and algorithms, was present in his "miraculous left hand". It is thus possible to see some "proto-version" of the art and science of mathematical modeling in Leonardo's way of work. Such modeling is a methodology of wide scope; in the discussion above on language and brain we showed how this method now extends to fields far beyond physics and engineering. With a sufficiently broad understanding of the key concepts structure, syntax and algorithm we have a method strong enough to allow for a common approach to knowledge of man and nature, to what there is and how we know.

References

Brown, C. M., & Hagoort, P. (Eds.). (1999). *The neurocognition of language*. Oxford: Oxford University Press.

Bursill-Hall, G. L. (1972). *Grammatica speculativa of Thomas of Erfurt*. London: Longman.

Capra, F. (2007). *The science of Leonardo*. New York: Anchor Books.

Chomsky, N. (1957). *Syntactic structures*. Den Hague: Mouton.

[17] See Ellis (2005) and Laland et al. (2010).

Clarke, D. M. (2003). *Descartes' theory of mind*. Oxford: Clarendon Press.
Dear, P. (1995). *Disciplines and experience. The mathematical way in the scientific revolution*. Chicago: University of Chicago Press.
Debus, A. G. (1978). *Man and nature in the renaissance*. Cambridge: Cambridge University Press.
Dijksterhuis, E. J. (1956). *Die Mechanisierung des Weltbildes*. Berlin: Springer.
Duhem, P. (1906–1913). *Etudes sur Leonardo de Vinci* (Vol. 3). Paris: Hermann.
Ellis, G. (2005). Physics, complexity and causality. *Nature, 435*.
Eloy, C. (2011). Leonardo's rule, self-similarity and wind-induced stresses in trees. *Physical Review Letters, 107*, 258101–258105.
Fenstad, J. E. (2010). *Grammar, geometry, and brain*. Chicago: University of Chicago Press.
Fenstad, J. E., & Wang, H. (2009). Thoralf Albert Skolem. *Handbook of the history of logic* (Vol. 5, pp. 127–194). Amsterdam: North-Holland.
Ferguson, E. S. (1993). *Engineering and the mind's eye*. Cambridge, Massachusetts: MIT Press.
Galluzzi, P. (1996). *Renaissance engineers. From Brunelleschi to Leonardo da Vinci*. Florence: Giunti.
Gärdenfors, P. (2000). *Conceptual spaces, the geometry of thought*. Cambridge, Massachusetts: MIT Press.
Harper, K. C. (2008). *Weather by the numbers, the genesis of modern meteorology*. Cambridge, Massachusetts: The MIT Press.
Hawking, S. (1980). *Is the end in sight for theoretical physics? An Inaugural Lecture*. Cambridge: Cambridge University Press.
Jackendoff, R. (2002). *Foundations of language*. Oxford: Oxford University Press.
Klein, S. (2010). *Leonardo's legacy*. Cambridge Mass: Da Capo Press.
Laland, K.N., Odling-Smee, J., & Myles, S. (2010). How culture shaped the human genome: Bringing genetics and the human sciences together. *Nature Reviews/Genetics, 11*.
Nicholl, C. (2005). *Leonardo da Vinci, the flights of the mind*. London: Penguin Books.
Paley, G. A. (1976). *Grammatical theory in Europe 1500–1700. The Latin tradition*. Cambridge: Cambridge University Press.
Pedersen, H. (1972). *The discovery of language. Linguistic science in the 19th century*. Indiana University Press.
Rocke, A. J. (2010). *Image and reality, Kekulé, Kopp and the scientific imagination*. Chicago: University of Chicago Press.
Rosch, E. (1978). Prototype classification and logical classification. In E. Scholnik (Ed.), *New trends in cognitive representations*. Lawrence Erlbaum Ass: Hillsdale NJ.
Snow, C. P. (1959). *The two cultures*. Cambridge: Cambridge University Press.
Turner, A. R. (1995). *Inventing leonardo, the anatomy of a legend*. London: Papermac.

Chapter 3
Relationships Between the Social and the Natural Sciences

Abstract An integrated science and technology policy is both complex and urgent. We have gradually come to understand that the relationship between science and technology is not neat and linear: it is not the case of first some basic sciences; thereafter some design and development; and finally the products, profits and the end to unemployment! A comprehensive science and technology policy is not a matter which is internal to science and technology, it also lies within the domain of several of the social sciences. This article reflects, in some generality, upon the relationship between the social and natural sciences, and examines some of the differences and similarities in methods and models as used by the social and the natural sciences.

Keywords Models in anthropology · Models in economics · Methodology and causal explanations · Taxonomy of mathematical models

It may be appropriate to approach to the topic in a somewhat roundabout way. Let us start by recalling the capital failure of a Norwegian project of the early 1950s which aimed to introduce modern science and technology to fishing in the province of Kerala in India. All the best that modern science and technology could offer was applied, but with few lasting effects on the local fishing community. This was, of course, not a unique failure, but it lead to NORAD (the Norwegian aid agency) at an early stage to ask the social sciences, in particular the anthropologists, for assistance in planning and operating development projects.

I do not claim that we, thereafter, were more successful in our aid efforts than others. There was always a tension between the engineers, who came with a technology in search of a market, and the anthropologists, who were trained to understand and possibly protect the values of the indigenous culture. Perhaps this is a tension which is still with us when we try to develop an integrated approach to science and technology policy?

My objective is to discuss in some generality the methodologies of the social and the natural sciences. However, I shall remain with the anthropologists for a while. They are supposed to be experts in bridging the gaps between different cultures, but there seems to be difficulties within their own field. Today the natural scientists, the biological researchers, search for evolutionary and physiological bases for social behavior, while the social scientists, the cultural investigators, are busy

J. E. Fenstad, *Structures and Algorithms*, Logic, Argumentation & Reasoning 15,
https://doi.org/10.1007/978-3-319-72974-9_3

deconstructing cultural texts and trends according to postmodern lights (see *Science*, vol 261, 1993). Perhaps the two cultures have separated and will never again meet? But, if anthropology splits, what remains of the general discussion of unitary methodologies!

However, let us not jump to conclusions. There is, of course, a difference in subject matter between the social and the biological anthropologists. But is there a similar and unbridgeable gap in the methodologies of the two cultures? This question is hotly debated—and is of some urgency. We are, for example, witnessing the development of two different, and almost disjoint, tracks in the training of graduate students in anthropology in the US.

I believe the situation is more complex than current fashion dictates. My starting point is an influential set of essays by the Norwegian social anthropologist Fredrik Barth on the role of models in social anthropology.

3.1 On the Use of Models in Anthropology

In 1981, a first volume of selected essays entitled *Process and Form in Social Life* by Fredrik Barth was published. One of the essays in the book bears the title '*Models*' *reconsidered*. Barth had, 15 years earlier, published a series of three very influential essays on the methodology of social anthropology. He now felt the need to add a fourth part. The original essays had attracted much attention and, as usual in scholarly intercourse, there had been much misunderstanding and criticism. Thus was a need 'to elaborate a few passages in what was no doubt too spare a text' (p. 76).

The starting point for Barth was the notion of *generative model*. This marked a shift away from the reigning structuralist description of that time in which forms, their structure and interrelationships, stood at the focus of attention. The new paradigm stressed a dynamic element and is, to a large extent, a model of how forms evolve or are generated. Some commentators have read too much into this claim criticizing Barth for attempting—and failing to give a story of creation, i.e. of how sociocultural phenomena are generated ab novo. Barth's focus was more on the *marginal changes* in form. His models are arrangements of data and facts intended to show 'how some facts affects the form of others, how some features of reality may be understood as the result, or epiphenomena, of other equally empirical circumstances and features—i.e. how some forms are "generated"' (p. 77). But rarely do we start from scratch, facts are, in most applications, situated with respect to pre-existing values and statuses; the changes of form that we observe, are in this respect marginal changes.

With the shift from the structural focus on frozen or rigid form to the dynamics of change, the notion of *process* emerge as central for the theory of generative models. Barth is very careful in restricting the rather loose and all-inclusive use of the word 'process'. For him a *process* is 'a generalizable set of linked events which keep recurring, the necessary interconnection of which, and the consequences of which, can be clearly described' (p. 78). In other words, what must be clearly described is the 'law' or rule giving the dynamic behavior of the model.

A résumé of a theoretical discussion tends to be 'doubly abstract' and far removed from experience and examples. It is therefore appropriate to emphasize that the theoretical writings of Barth are firmly rooted in an extensive practice of anthropological field work. His theoretical concepts are, in their substance, extracted from this experience and have been tested and refined by further studies in the field. One type of process that has come from this work is captured by his notion of transaction, which in typical cases describes a process where certain actors or groups of actors arrange their mutual dealings with each other in ways which protect and respect their rights and assets; an example studied by Barth is the transactions within the network of skipper-netboss-fishermen, where a clear conception of self-interested reciprocity in protecting the fund of rights and assets of the different actors within this status set provides the 'dynamics' of the process; i.e. determines how they act. It is not, therefore, unexpected that an 'emergent' property of a transaction is that the value the participants, whether as individuals or groups, ascribe to the objects or goods of the transaction tends to become more congruent.

Another important part of the theoretical scheme of Barth is his distinction between two levels in a process, the 'event' versus the 'aggregate'. For Barth these are two different levels of the same reality: 'I understand the constituent event of social life to be characteristically micro, performed by actors (individuals or corporate groups). These events or acts are conditioned by, and in turn together constitute, the aggregate level. Processes provide the key conceptualization for depicting how aggregation comes about, and explaining aggregate form. The aggregate has markedly emergent properties, ...' (p. 80).

An example was mentioned above, the transactions within the network of skipper-netboss-fishermen. The micro level, in which the dynamics are defined, is constituted by the skipper, the netboss and the fishermen, i.e. the actors. The aggregate or the macro level, is the single status set of skipper-netboss-fishermen; as noted above, an emergent property of this aggregate, is the congruence of values of the diverse actors.

A final ingredient of the theoretical scheme is the insistence of Barth 'to unite in one single analytical model what has variously been separated as culture versus society, meaning versus instrumentality, saying versus doing' (p. 81).

This does not necessitate any major revision of the analytical scheme. It means that when one describes the dynamics of a process one must bear in mind that an act can be both *described*, i.e. be an external observable, and *understood*, i.e. carry an intentionality or meaning. None of these aspects can be eliminated, nor can one be reduced to the other, e.g. an act may have material consequences independent of the actor's intention, which means that we cannot eliminate the observational description of acts.

This completes the description of the model. There are a number of remarks to add. The notion of *power* has been absent from the discussion. This is deliberate from Barth's side. A concept of power is, at most, derivative. Who has 'power' in the complex transactions in the status set skipper-netboss-fishermen? The key notion is 'choice', leading to decisions and acts, the very ingredients which drive the dynamics of the process.

The concept of 'value' is, however, central. But in Barth's conceptual scheme it has nothing to do with ideas of rightness and morality. For Barth value is related to the dynamics of a process: 'Values in this sense are thus synthetic accounts of an actor's or a group of actors' codification and ranking of a diversity of ends and goals, where the aspect of ranking is embodied in a praxis of choice. Values thus serve to identify, within a particular culture, the purposes and priorities which inform an actor's choices and organize his efforts' (p. 95).

And further: 'Analysis in terms of such a concept of value thus achieves a synthesis of micro- and macro-levels which a normative perspective cannot provide' (p. 96).

The modern reader, familiar with at least some popular account of games, decisions, chaos, and non-linearity, may mistakenly believe that we here see another example of a band-wagon phenomenon, the urge to be fashionable by dressing up your 'science' in the latest buzz-words of the marketplace. Any such reader should carefully note the date of the essays. The first group dates from the mid 1960s, the fourth from 1981. A reader of Barth's work will be struck by the rich context of observation and field research that the theoretical scheme is set into, much of this work is due to Barth himself. Barth wants to shift the paradigm from a static structural description to one incorporating change and the emergence of form and structure. Barth's father was an eminent mineralogist with a sure command of the physics and chemistry of his time. Barth entered the scene when game theory first made its impact on the social sciences. Influences could probably be traced; here it suffices to note that statistical mechanics, dynamical system theory and game theory provide a rich arsenal of concepts, and that we see clear and sometimes explicitly acknowledged relationships between those sciences and elements of Barth's model; e.g. the relationship of event, aggregate and the emergence of form to concepts in statistical mechanics and thermodynamics, and the possible use of ideas of ideas and techniques from game theory in the maximization of values and the subsequent choice of acts. But we do not see any spurious formalizations and import of extraneous mathematics into the model; the analysis is definitely driven by the facts and data of anthropology.

3.2 A Digression on General Methodology

At this point it may be useful to preface a fuller discussion of models and methodologies in the social and natural sciences by a brief review of some current themes in the philosophy of science. There is almost always a time-lag between the current theoretical and methodological discussion within a particular field of study and the way this field is presented in academic teaching and general discussion; I except some current trends in science journalism. In most cases this is of little importance, there is a hard and stable core which form the nucleus of any serious curriculum. But stability can sometimes be carried too far. For too long has the public view of the methodology of science been dominated by the 'hard-liners' of the hypothetic-deductive method, such as Carnap, Popper, Reichenbach and Hempel—the authors of the canonical text, a prime example being Hempel's *Philosophy of Natural Sciences* from 1966—in which the simple case of mechanics (both classical, relativistic and quantum) and a few well-chosen cases from medical research set the stage.

There has been an added dynamic and historical dimension to this body of teaching through the work of Kuhn, Lakatos and others, but the impeccability of the triad: (i) observation/experiment; (ii) conjectures/theories; and (iii) predictions/refutations/applications, persists and is used to explain how conjectures and theories arise out of data, how applications and predictions are deductively derived from theories, and how further data, either confirm or refute our predictions, hence also our conjectures and theories. The process repeats itself, and therein a progress towards ever greater insight and understanding. Of course, science—in particular, applications of science—have their problematic sides. We therefore tend to shy away from such words as 'progress' and instead stress that understanding needs to be coupled to concerns with themes of responsibility and of the ethics of science. But that is not the theme of this essay.

Against the orthodoxy of academic teaching I would like to draw attention to two books by Nancy Cartwright, a philosopher of science at the London School of Economics. The books are: *How the Lawes of Physics Lie* and *Nature's Capacities and their Measurement.* Cartwright argues against the standard hypothetic-deductive picture of science. According to this view, laws of nature are functional; typically, they consist of a set of equations describing how a system works, e.g. a set of 'particles' traces out a trajectory in a suitable space over a given period of time. We have forces, velocities, masses, and trajectories, but *not* causes. This is part of our heritage from Hume, and it has had a profound influence on our view of nature and explanation. It is this view which is being challenged.

The book *Nature's Capacities* opens with three theses: science is measurement; capacities can be measured; and science cannot be understood without them. Thus, measurements and causal explanations are at the core of Cartwright's philosophy of science. In her first book *How the Lies of Physics Lie* she argues that doing science in practice consists of unraveling causal relationships. She is not the only current philosopher of science who argues in this way. In fact, the hypothetic-deductive methodology and the standard view of explanation through so-called 'covering laws' has been effectively criticized from many quarters. I cannot in this essay do justice to this discussion; at most I can add a few observations. Doing this I run the risk of misrepresenting the view of the different participants. Let me, therefore, accept responsibility for the observations I am going to make; my excuse is that they will be of some relevance for later part of this chapter.

The first observation is that the laws of physics never truthfully apply to nature. My favorite example is boiling water; for which we have impressive theories of statistical mechanics and phase transitions. The latter are connected to singularities in the so-called partition function of the system. In finite domains the partition function is analytic and therefore has no singularities; they can only appear when we pass to the infinite thermodynamic limit. However, any amount of hot water, in nature, occupies a finite volume. So how can water boil?

This is not the only example; try, for example, to deduce the behavior—the true behavior—of a real laser from the standard equations of textbook exposition. You will find that there are quantum leaps filled with specific ad hoc tricks. The law-believer is, however, not likely to admit defeat, but will argue that laws are idealizations and that in every application some corrective terms are required. But the engineer's feel for materials in the actual construction of a laser goes beyond the defense through corrective terms. In fact, much of engineering goes beyond text-book doctrines of scientific explanations; see for example the book by E. S. Ferguson *Engineering and the Mind's Eye*.

There has also been a greater appreciation in the philosophy of science for the complexity of experiments with implications for our understanding of what are the 'hard' data underlying theoretical explanation (see the book by I. Hacking, *Representing and Intervening*). This has been evident in the setting of large-scale experiments in high energy physics, but occurs everywhere. The interpretation of a PET (positron emission tomography) scan of the brain location of some cognitive task depends upon non-trivial physics and extensive mathematical modeling, e.g. raw data have to be transferred to a standard brain format and enhanced through a delicate center-of-mass algorithm before they become the reported data in the analysis of the cognitive task.

This leads back to science seen as a search for causal explanations. In fact, much of everyday science is concerned with the causal unravelling of complex phenomena where experiments and explanation are intertwined in countless ways. Laws—in their splendid abstractions—have a role to play. In most cases, these general laws are pieces of mathematical analysis developed, not on slices of reality, but in the context of highly abstract (set theoretic) structures. From this point of view, laws are tools which are at our disposal in constructing causal explanations and computing predictions; sometimes this can be done in a remarkably comprehensive and consistent manner, as for example in the applications of classical mechanics to space travel; sometimes it is blatantly ad hoc, as for example Bohr's explanation of spectral lines. Laws are good for calculations; the infinite thermodynamic limit is a fiction, but good for explicit calculations. Laws are tools which we may or may not choose to use; with this understanding we may legitimately ponder whether we shall or shall not include a relativistic correction—the one is not in an absolute sense more true to nature than the other.

Laws in the above sense describe highly idealized situations. However, they are used in the understanding of messy phenomena. This brings us back to the three theses of Nancy Cartwright. It is true that we develop our computational schemes in highly idealized situations, e.g. we start our study of hydrodynamics by setting down equations for an ideal fluid flowing between parallel banks and with no inner viscosity. How is it possible that these insights apply across a variety of situations? How can the 'laws' of hydrodynamics tell us anything about the flow of dirty water in an actual river? Cartwright would say that it is because science is concerned with the study of capacities, i.e. of stable and enduring properties or tendencies of causes to bring about specific effects. Causal explanations have been banished from the methodology of the natural sciences, but not necessarily from the sciences themselves. There is also a tradition of causal modeling in econometrics and several of the other social sciences. In the last century J. S. Mills was a champion of causality and tendencies—or capacities in the present terminology. Cartwright explicitly acknowledges the insights of Mill, but her discussion proceeds far beyond his discussion of the Moral Sciences of economics and politics. Not everyone will agree with the three theses: science is measurement; capacities can be measured; science cannot be understood without them. I am to some extent in agreement with her view on causality and laws, but is all of science measurement?

3.3 A Taxonomy of Mathematical Models

In this section I will take a closer look at the use of mathematics in modeling and understanding. This is a vast topic and too large for this discussion. I will therefore choose one particular perspective, which in some ways continues our discussion of the use of models in anthropology. The class of models discussed in this section is not at all uncommon in the social and natural sciences; we are often interested in the development or behavior of some population over a period of time. These models can be classified according to the dynamics which governs the process.

3.3.1 Mechanical Man and Nature

This way of doing science represents our Newtonian or Cartesian heritage. One or an ensemble of particles is subjected to known forces. A law, always represented by a linear algebraic or linear differential equation, is given. When parameter values are set, the future development of the system is uniquely determined. At most an initial mover is needed, the rest is determined. This is the point of view of Laplace, and the mathematics is the theory of unitary groups, symmetric with respect to time reversal.

Such is, in extreme brevity, the mathematics of classical physics, of both mechanical and electromagnetic phenomena. It is also the mathematics that is in deep harmony with the orthodox views on the hypothetic-deductive methodology discussed above. It is, moreover, the mathematics of industrial applications and traditional large-scale scientific computations.

In the latter case we assume that our equations faithfully describe a piece of reality, which can either be a natural phenomenon or some manmade artifact. From our equations we extract algorithms—and then we compute. The result can either be predictions about the development of some physical system or a set of specifications for some construction task. Our trust in the equations can sometimes be so great that we declare the end results of the calculations to be 'true facts of nature'. Thus, in some areas of science we have seen a 'computational' sub-discipline added to the traditional division into the theoretical and the experimental parts. The added computational part is sometimes used to extend the result to parameter values beyond those obtainable though actual experiments and observations, and sometimes used to speed up the time development of some process. In industrial applications we sometimes see computations replace actual tests of designs, for example the reliance on computer simulated collisions instead of the physical destruction of cars.

The strict orthodoxy of mechanical man and nature will, however, have to be modified in various ways.

3.3.2 Adding Chance and Uncertainties

In the above we assumed precise mechanisms and measurements, i.e. laws were deterministic and initial values could be exactly set. However, impreciseness can be added in several ways.

(a) We are not able to measure initial values with preciseness, either as a practical concern or as a matter of principle. The former case adds nothing new in principle; there is a well-developed 'theory of errors' and many techniques of statistics to take care of this kind of impreciseness or 'uncertainty'. The latter case, where uncertainty enters as a matter of principle, adds a new dimension. This is the case of, for example, quantum mechanics, where the input is a probability distribution, but the process—in this case governed by the Schrödinger equation—is still given by the mathematics of the unitary group.

(b) In the second case, the law is no longer exact but incorporates randomness in various ways. This is the domain of stochastic equations, where the unitary group is replaced by a Markov semigroup, which means that the process is time irreversible (the semigroup property) and that the future is determined by the past only through the present (the Markov property).

The use of stochastic equations, both linear and differential, has seen a significant increase within mathematical modeling. The approach can sometimes be seen as somewhat opportunistic—in the absence of detailed knowledge of a phenomenon, one can alway add a random term! But in many cases our insight into the probabilistic laws is precise and the domain of application is wide, as for example seen in the somewhat esoteric case of 'an application of reflected diffusion to the problem of choosing between hydro and thermal power generation'.

Common to the types of models discussed so far is a belief in measurements and numbers. It is told of the theoretical physicist J. Schwinger that, during the war, he was much appreciated by his experimental colleagues by returning with numbers when asked for assistance, not with general and computationally intractable formulas. Further, as some physicists like to tell, it was the exact measurements and numbers associated with the Lamb shift which drove the theoretical development of quantum field theory in the period 1946–1951 (see Schweber 1994). This attitude is well captured by the following quote: 'All the young theoreticians at the wartime laboratories learned that physics is about numbers and about the results of experiments.

Good theories yield numbers, explain numbers, and help design good apparatus. They allow you to control and manipulate the objects they describe' (Schweber, p. 144).

3.3.3 Chaos and Catastrophes

The unitary group is the paradigm of linearity and the ability of exact prediction. But despite centuries of physics and technology there are, in fact, few natural or social phenomena that can be usefully be modeled by this tool. One way of adding 'change and uncertainties' was discussed above. Another way is to continue to use deterministic equations, but to extend the study of equations from the linear to the non-linear domain.

Non-linearity, which often occurs in cases where there is a sensitive dependence on initial data, was always recognized as part of physical theory; one prominent example from the late 19th century is Poincaré's study of the stability of the solar system (i.e. the long time stability of a deterministically given many-body mechanical system). However, despite the genius and the efforts of Poincaré, the problem resisted analytical treatment; the appropriate mathematical tools were simply not there.

The development of the theory of dynamical systems from Poincaré to the present day cannot be told here. The advances have been impressive and we have today a reasonably mature mathematical theory. But one of the ironies of the history of non-linearity is that the first convincing modern application to the natural sciences was made by a person outside this line of development. In 1963, E. Lorenz, a meteorologist, published a paper *Deterministic Non-periodic Flow*, which had all the right insights. His equations were a simplified system of the equations used for weather forecasting, but complex enough in their non-linearity to capture the 'chaotic' or unpredictable development of weather systems.

Another early application proceeded in a methodologically 'correct' way, i.e. as an application of the mathematical theory. This was the study initiated by D. Ruelle on the onset of turbulence. Hydrodynamics is a difficult subject. The equations are assumed to give a complete description of the phenomena, but in distinction to celestial mechanics it is difficult to unravel their 'secrets'; turbulence, in particular, has been difficult to understand. At one time random elements were added to understand the onset of turbulence; this would be to follow the second strategy of the previous section. The insight of Ruelle was that the chaotic behavior of non-linear deterministic equations already contained the complexity needed (i.e. strange attractors) to understand the onset of turbulence.

Chaotic behavior, strange attractors, the emergence and stability of forms, and the inability of long term prediction are some of the 'buzz-words' of contemporary non-linear science. I shall not give any further explanations here: both the serious and the popular literature is vast. I recommend to the curious who want an authentic account, two books, one by Lorenz, *The Essence of Chaos*, and the other by Ruelle, *Chance and Chaos*. Despite somewhat fashionable titles, this is the story told with insight by two of the originators.

Let me conclude this section by some remarks on prediction and computations. With non-linearity we are witnessing a shift of paradigm. The deterministic laws of classical physics lead to exact predictability; computational techniques were developed to exploit this feature; it was possible to describe in quantitative details either the time development of a physical system or the specifications of some industrial product. This ability was also matched by the new technology of large-scale computers. With non-linearity there is no guarantee of long-range predictions. This was the conclusion Lorenz drew in 1963 from his studies.

However, instead of computing a unique future we can simulate possible developments; large-scale computing can be made to serve a different master. Thus, there has been a shift or, more correctly, an extension from a 'hard' science of 'computable' prediction to a broader science or art founded on insight; we cannot for example compute the exact time for the onset of the next major earthquake in California; but we can use our model to gain insight into the phenomena and to develop strategies against adverse effects.

Extending basic physics to weather forecasting and earthquake prediction has been a useful lesson for the hard-liners of scientific methodology. Do we also, in these methodological developments, see emerging similarities between the social and the natural sciences? Before attending to that question I shall make some remarks on the sometimes turbulent relationship between mathematics and economics.

3.4 The Case of Economics

In the 19th century terminology of J. S. Mill, economics was one of the Moral Sciences. But it was a science, not a branch of philosophy and, in the view of Mill and many others, destined to become an equal partner in the effort for the betterment of mankind. But, if it was to become an equal partner, must it not also be written in the language of science, i.e. in symbols and numbers?

A Newtonian heritage may sometimes be a burden. Immediately following Newton's *Principia Mathematica* several curious examples of 'scientific respectability' arose. My favorite is from 1699, *Theologiae Chrstianae Principia Mathematica*, by J. Craig; a treatise developing 'theorems' such as 'Velocities of Suspicion produced in equal periods of time increase in Arithmetical Progressions'!

Such examples may amuse. It is, however, more important to note that the emerging 'moral sciences' presented problems and challenges which have had great influences on the further development of several of the mathematical sciences, in particular, on developments in probability, statistics, and geometric algebra. But in some of the emerging social sciences we saw a 'physics envy', a self-conscious awareness of not being quite good enough. This has lead to a certain amount of 'physics transfer', i.e. of attempts to cast theories in the mould of the reigning fashion in physics. Let us consider some examples of physics transfer in economics and relate them to the taxonomy of mathematical models.

The first example is related to the mathematics of the unitary group, i.e. to the mathematical paradigm of the physics of the last part of the 19th century. The example from economics is the neoclassical theory of the 1870s, associated with the names of Jevon, Marshall, Menger and Walras. The mathematics of this example is the theory of vector fields and the setting is demand theory where the price vector is the gradient vector of a utility potential; the task is one of optimizing under suitable side conditions. One can argue whether this was the paradigm for all neoclassical economics; one can further argue how well the mathematics and physics was understood among the pioneers; one cannot dispute the fact that this was an important example of 'the successful penetration of mathematical discourse into economic theory'. There is a lively current debate concerning these issues—the interested reader is referred to the book *More Heat than Light* (1989) by P. Mirowski and the recent collection of essays *Non-Natural Social Science: Reflecting on the Enterprise of More Heat than Light* (1993).

The influence of the field paradigm has remained strong; not only as a research program internal to economics; it has also been strengthened by the many links between the physics and economics community. One example is the case of the young J. Tinbergen. There is a direct link from L. Boltzmann, one of the great 19th century physicists, via his student P. Ehrenfest who, in 1929, was the thesis supervisor of Tinbergen; the thesis topic was *Minimumproblemen in de natuurkunde en de economie*, an explicit example of physics transfer. Another link can be seen from J. Gibbs, of equal eminence as a late-19th century physicist, via E. B. Wilson to P. Samuelson, whose *Foundation of Economics* from 1947 is very much within the neoclassical paradigm.

The mathematics of economics is, however, of more substance than simple-minded 'physics envy'. Our next example is taken from the second type of model discussed in the previous section. The tool this time is linear equations, and the goal is causal modeling. This is a paradigm associated with the later Tinbergen and R. Frisch. The, perhaps, decisive contribution was a seminal paper by T. Haavelmo, *The Probability Approach in Econometrics* (1944), which reported work done during his tenure at the Cowles Commission. A typical example is again to determine how demand depends upon price. The link is supposed to be linear, but the constant term of the linear equation is now taken to represent a random influence, turning the equation from a deterministic one to a random one. The important task is to estimate, i.e. to find some numerical expression for the coefficient of the price variable. The stability of this coefficient across random environments represent the tendency of price to influence demand. There is an insightful discussion of this model in Cartwright's *Nature's Capacities and they Measurement*. Obviously, for Cartwright the tendency of price to influence demand is a capacity, pure and simple, to be measured and to be used in causal explanations.

It is not my intention in this section to present a complete survey of mathematical models used used in economics. The work in Haavelmo was an example of work driven by internal demands; it was not an example of transfer. I should add the remark that 'physics transfer' is not in itself a methodological mistake; but problems can arise when critical assumptions become unnoticed or ignored in the transfer.

I shall mention two examples connected with symmetry. The first is related to the economics example above; for the price to be the gradient of a utility vector field, the associated Jacobian matrix must be symmetric; this is often an unnoticed assumption. The second comes from cognitive science; in order to use the mean-field theory of spin glasses to derive quantitative results about brain processes, the assumption is routinely made that the interaction between sites are symmetric, ignoring the fact that this is rarely true of the interaction of neurons.

There are cases where the transferred model has later, to some extent, been 'internalized'. One example from economics is the Arrow-Debreu theory of equilibrium, where methods from algebra, geometry and fixed-point theory replace the vector analysis of the neoclassical approach. But abstraction and purification may raise other problems: has the Arrow-Debreu theory in its generalized version the power to generate causal explanations in specific applications, or has it been turned into an 'empty' explanatory shell, i.e. a metaphor; see the discussion in B. Ingrao and G. Israel, *The Invisible Hand* (1990).

The final example will be related to models that are connected with non-linear dynamical processes. The flavor of this example is well expressed through the title of the collection of essays coming out of an early conference, *The Economy as an Evolving Complex System* (1988). Here, we see a straightforward case of physics transfer. Ruelle, who was a participant in the meeting, has a short section on economics in his book *Change and Chaos*.

'At a low level of technological development, the economy should have a steady-state corresponding to the steady state of a fluid layer weakly heated from below … At higher levels … we expect that periodic oscillations may occur. In fact, economic cycles, also called business cycles, have been observed and are roughly periodic. At still higher technological levels, we might see superpositions of two or three more different periodicities, and economic analysts have indeed seen such things. Finally, at sufficiently high levels … we may have a turbulent economy, with irregular variations and sensitive dependence on initial conditions. One may argue that we now live in such an economy' (Ruelle, pp. 82–83).

Ruelle is no naive trend-follower, he has an expertise in applying mathematics and is known for his contributions to hydrodynamics and statistical mechanics; in this case he wrote tongue in cheek, well aware of the difficulties of turning a metaphor into testable science. Not everyone shows equally good taste: we have seen much opportunistic 'science' in the wake of chaos, catastrophe and non-linearity. But, if part of this metaphor can be turned into a viable tool, we will possess an extended scheme for causal explanations. The scheme will, however, not necessarily be 'numerical' in the traditional sense of physics; it may, however, in some cases be an equal useful guide for planning and action.

3.5 One or Several Sciences?

How do the social sciences fit into this scheme? Economics has had its share of the discussion. The case of social anthropology is interesting and for many, perhaps, unexpected. The research program developed by F. Barth and briefly reported above, fits well with the views set forth in this chapter. There certainly is a common ground in the search for structure and explanation when anthropology is included; at least if social anthropology is more than deconstruction of cultural texts in a postmodern light (see the reference to *Science* above).

The other social sciences are supposed to lie somewhere between economics and anthropology on any reasonable scale of measurement; a simple case of interpolation should therefore suffice to close the discussion. This is, perhaps, the proper view to take. We no longer insist on 'computable predictions' as the defining characteristics of doing science, even within the traditionally 'hard' sciences. This does not imply that we are abandoning every notion of measurement. A case can be made that science is measurement, not story-telling.

The art and science of measurement is complex and the scales used to measure are many. The literature on this topic is vast and cannot be reviewed here; but are comprehensively covered in the *Foundation of Measurements* by Krantz et al. I shall remind the reader of only one simple point: in processes governed by a unitary group we have an extreme case of the ability to measure and predict (as in the case of the Lamb shift); this is an important and, perhaps, an enviable case, but it is not the defining case of science.

Turning from measurements to models, we have seen that there are a number of approaches that have given or promise to give interesting applications in the biological and cognitive sciences. Not everyone may agree with this, for example, the neurophysiologists are, as a general rule, suspicious of the use of 'mathematical' models. But this may be a case where they more correctly should have been critical of the premature, naive or plainly wrong use of such models. There is no logic which leads from the incorrect use of models to the non-existence of 'correct' models. In the cognitive sciences, we see today a research program where the aim is to link the phenomenology of the mind with the concrete dynamics of the brain. It is no wonder that the established tribes of anatomists and physiologists are skeptical.

Our adopted perspective in discussion of causal explanations was the model of how systems develop over time. Looking more carefully at the mathematics of this modeling process we observe that two extreme cases seem amenable to numerical treatment. 'Small' ensembles, e.g. one particle, or systems of particles, which can be modeled as one-particle systems, and 'large' ensembles, i.e. systems which can be modeled by a large, even an infinite, number of particles. Some models from cognitive science and economics fall into these categories. But several of the social sciences are characterized by populations of 'medium' size. For such systems there seemed to be no adequate mathematical tools. The combinatorial analysis which is so powerful for small systems, becomes intractable (even on the largest computers), and the analytical tools (differential and integral calculus) of large systems are inapplicable.

Game theory was originally thought to be the correct tool in the medium-sized situations, but at present the theory is either 'combinatorial' and cannot cope with medium-sized populations, or it is 'differential' and need s assumptions of large populations. Lacking an explicit mathematical analysis, i.e. 'closed' formulas, we have in such cases witnessed an explosion in simulation studies made possible by recent developments in computing technology. But even the largest computers have not to this time been able to cope with the exponential growth which, as a rule, is associated with the iteration of choice alternatives in decision theoretic analysis. This situation may now start to change, see the discussion in the introductory chapter on 'big data' and computational feasibility.

In concluding, I want to return to the introductory remarks where I commented on the contrast, even tension, that was sometimes seen in development projects between, on the one side, the engineer in search of a market and committed to growth and, on the other side, the anthropologist trained to understand and protect the values of an indigenous culture. I even speculated that this is a tension which is still with us when we try to develop an integrated approach to science and technology policy. In many ways we are still an 'indigenous culture' in need of an improved policy for 'sustainable development'. This is not an easy goal to reach; a possible tension in methodology and aims could turn into an effective barrier against a joint development of an integrated policy. This is the negative conclusion we would have to draw if the current debate within American anthropology is a paradigmatic case.

Contrary to this I have tried to argue in this essay that, seen in a broader scientific, technological and humanistic perspective, there is a common ground in the search for structure and explanation: we have seen a coherence and a continuity across disciplinary boundaries in our attempt to understand man and nature. Thus, to be brief, it is not the incompatibility of methodologies which is the barrier—if one exists—to a joint undertaking of the natural and social sciences towards the development of an integrated and sustainable future policy.

Advice on Literature

The discussion in this essay has touch upon many different topics. A complete set of references would be lengthy and probably not of interest to the general reader; some documentation of sources is, however, necessary and some advice on the literature may, perhaps, be of interest to the general reader.

Philosophy and methodology of science is an established academic discipline with a vast textbook literature. Most readers will have been exposed to an ample dose of the hypothetic-deductive method. In this essay I have relied on the well recognized work of Nancy Cartwright to present a picture more attuned to the actual practice of science:

Cartwright, N. (1983). *How the laws of physics lie*. UK: Clarendon Press.

Cartwright, N. (1989). *Nature's capacities and their measurement*. UK: Clarendon Press.

Another contribution which, in an interesting way, complements Cartwright's discussion on the role of experiments and the existence of theoretical entities, is:

Hacking, I. (1983). *Representing and intervening*. Cambridge: Cambridge University Press.

We have also seen a remarkable growth in the studies on the history of science. One example, which explicitly acknowledges the influence of Cartwright and Hacking, and which I have quoted above, is the recent history of quantum electrodynamics:

Schweber, S. S. (1994). *QED and the men who made it*. New Jersey: Princeton University Press.

Technology is often seen as an application of science, but technology has its own history and methodology; one reference is:

Ferguson, E. S. (1993). *Engineering and the mind's eye*. US: MIT Press.

Science is measurement, but measurements are more than numbers produced in high energy physics; for a general discussion see:

Krantz, D. H., Luce, R. D., Suppes, P., & Tversky, A. (1971, 1989). *Foundation of measurement*, Vol. I–III. Cambridge: Academic Press.

The current situation in American anthropology is described in two articles in Science, Vol. **261**, 24 September and 10 December, 1993. The discussion of models in anthropology is taken from:

Barth, F. (1981). *Process and form in social life*, selected essays, Vol. I. UK: Routledge and Kegan Paul.

There is a vast textbook literature on mathematical modeling; I will therefore not give further references. In the text I recommended two basic sources on non-linear themes:

Lorenz, E. N. (1993). *The essence of chaos*. Washington: University of Washington Press.

Ruelle, D. (1991). *Changes and chaos*. New Jersey: Princeton University Press.

Some may want to consult the writings of I. Prigogine on these topics:

Prigogine, I. (1980). From Being to Becoming, W. H. Freeman.

Discussing economics I made a number of references: neoclassical theory and physics transfer; causal modeling and the proper use of probability; general equilibrium theory and its empirical content; economics and chaos. The sources are:

Anderson, P. W., Arrow K. J., & Pines, D. (Ed.). (1988). The economy as an evolving complex system. US: Addison-Wesley.

de Marchi, N. (Ed.). (1993). *Non-natural social science: reflecting on the enterprise of more heat than light*. US: Duke University Press.

Haavelmo, T. (1944). The probability approach in econometrics. *Econometrica, 12*.

Ingram, B., & Israel, G. (1990). The invisible hand. US: MIT Press.

Mirowski, P. (1989). *More heat than light*. Cambridge: Cambridge University Press.

Chapter 4
Changes of the Knowledge System and Their Implication for the Formative Stage of Scholars: Experiences in the Natural Sciences

Abstract In this chapter we review some recent trends in the natural and biomedical sciences related to issues of complexity and reductionism, mathematical modeling and simulations, with particular reference to system biology, and discuss the impact of these developments on future research training.

Keywords Science and ignorance · Complexity—reductionism
System biology · Research training

4.1 Science Is Shaped by Ignorance

When the journal *Science* celebrated its 125th anniversary, it did not do so by pointing in pride to the enormous advances in science and to the many applications in the 125 years the journal has existed. Instead it chose to focus on what we don't know by asking 125 "hard" questions, one for each year of its existence, about what we do not know.[1]

Science is, as we all agree, a cumulative enterprise. In Newton's words, "we are standing on the shoulders of the giants of the past." But we must be aware of the danger of becoming the mere custodians of our heritage and the tame cultivators of well-known territories. We must always move beyond, crossing the borderline between knowledge and ignorance by asking difficult—and sometimes awkward—questions that science cannot yet answer. In the words of David Gross, the 2004 Nobel physics laureate, "science is shaped by ignorance".

But not every question of ignorant men makes good science. Science consists in doing what is do-able. The art of science is asking the "right" question. Let, therefore, ignorance and good taste be the chosen context for our remarks on the formative years of scholars.

[1] *Science* vol. 309, 1 July 2005.

J. E. Fenstad, *Structures and Algorithms*, Logic, Argumentation & Reasoning 15,
https://doi.org/10.1007/978-3-319-72974-9_4

The list of questions in *Science* is a useful guide to *ignorance*. The list is divided in two parts, the top 25 and a follow-up of 100 supplementary questions. Let me concentrate on the top 25 to get an impression of where the main domains of ignorance lie. The list can somewhat arbitrarily be divided into four parts:

i. physics and cosmology—4 questions;
ii. computing, chemistry and the earth science—3 questions;
iii. environmental sciences—3 questions;
iv. biology and health—15 questions.

The first sublist is fairly homogeneous and asks what is the universe made of, can the law of physics be unified, are we alone in the universe, and do deeper principles underlie quantum uncertainty. The next group is notable for its brevity and asks one question from each of the mentioned fields, what are the limits of computing, how far can we push chemical self-assembly, and how does the earth's interior work. The third group is equally brief and asks how hot will the greenhouse world be, what can replace cheap oil, and will Malthus continue to be wrong. These lists add up to 10 of the 25 questions. The remaining 15 questions come from biology and the health sciences and cover a wide area ranging from molecular biology, through system biology, to health issues. Some examples are what is the biological basis of consciousness, why do humans have so few genes, how can a skin cell become a nerve cell, how and where did life on earth arise, what determines species diversity, how will big pictures emerge from a sea of biological data, and is an effective HIV vaccine feasible.

One could argue at length about each individual question on the *Science* list, whether it belongs among the top 25 or not. What is noteworthy is the distribution over sub-lists. This is the choice made by *Science* and it presents a clear picture of where our ignorance lies. At the same time there is also an implicit priority in the list. I grew up in an era when the physicist's equivalent of General Motors (GM) chairman Wilson's dictum about GM and the USA was the valid paradigm—what is good for physics is good for science—and priorities were set accordingly. If we believe in the distribution from the *Sciences* list, with 18 from environment, biology and health, we should instead adapt a famous saying of President Kennedy, and ask not what is good for physics, but ask what can physics do for biology.

Becoming aware of the unknown is part of the education of the young scientist. Being prepared for the unknown is another matter. Facts and tools are the basic ingredients, but talent and insight—which was meant by using the words *good taste* above—are needed to ask the important, but do-able questions. Facts and tools are to a large extent text-book material and can be learned by almost every one. But almost always the results of such activity are at most useful, but rarely ground-breaking science. No one can yet offer a convincing analysis of what insight really is. We know that in our context it is something you develop by doing science, a kind of "implicit" skill you acquire by working on "hard" problems and which allows you to ascend beyond mere facts and to see and explore patterns and structures emerging from "the sea of observational and experimental data", as stated in the list of questions from *Science*. Today "hard questions" almost always require that one crosses traditional disciplinary boundaries. *Good taste* accordingly demands insight from many fields. And since good taste and insight are the result of doing science, it presupposes that the young scientists in their formative years do science in the right kind of environment, a point we shall return to in the last section of this paper.

4.2 Beyond Reductionism

In 1980 Stephen Hawking gave his Inaugural Lecture as Lucasian Professor of Mathematics in the University of Cambridge. The title was *Is the end in sight for theoretical physics?*[2] Let me quote the opening statement: "In this lecture I want to discuss the possibility that the goal of theoretical physics might be achieved in the not too distant future, say, by the end of the century. By this I mean that we might have a complete, consistent and unified theory of physical interactions which would describe all possible observations." He notes previous attempt at a final synthesis and their failures, but this time he is cautiously optimistic. There is a strong undercurrent of reductionism in his text. He seems to assert that if we have the final equations of physics, then everything else will follow. He notes Dirac's reduction of chemistry to physics, but is careful to explain that this is only "in principle". Real systems are too complex for detailed calculations and one has "to resort to approximations and intuitive guesses of doubtful validity". He notes with some regret that "although in principle we know the equations that govern the whole of biology, we have not been able to reduce the study of human behavior to a branch of applied mathematics".

Reductionism has been a strong guiding principle in scientific research. We understand the complex by first analyzing the parts; thus complex behavior is a synthesis of the behavior of individual component parts. For a powerful example of this strategy see the recent review article on computational quantum chemistry, *From atoms to crystals: a mathematical journey* (Le Bris and Lions 2005). I have chosen this example for two reasons; it has relevance to the quote from Hawking above, and it is a good example of what serious work on reductionism means.

Despite the heroic efforts of computational chemistry, doubts about the reductionist strategy are heard more and more. A noteworthy example of this trend is a survey of complex systems in *Science* a few years ago.[3] The journal has consciously chosen "card-carrying" scientists to explore these themes, with examples from current work in chemistry, biological signaling systems, the study of the nervous system, evolutionary trade-offs in animal aggregation, natural landform patterns, climate research, and the economy. Much has happened since 1999, but in my opinion the *Science* survey still offers a balanced overview of complex behavior in non-linear systems. Let me just add another fine specimen of applied non-linear science, a review article from *Nature* (Scheffer et al. 2001) on *Catastrophic shifts in ecosystems*. In this article

[2]Hawking, S.

[3]*Science*, vol. 284, 2 April 1999.

a concept of *resilience* is explained in terms of some simple examples of non-linear dynamical systems, and a case is made that strategies for sustainable management of such systems should focus on maintaining such resilience. As the authors note in the introduction to their review: "All ecosystems are exposed to gradual changes in climate, nutrient loading, habitat fragmentation, or biotic exploitation. Nature is usually assumed to respond to gradual change in a smooth way. However, studies on lakes, coral reefs, oceans, forests and arid land have shown that smooth changes can be interrupted by sudden drastic switches to a contrasting state." These abrupt changes can in many cases be explained by assuming that the systems are driven by a non-linear dynamics. Shifts in such systems cannot be predicted in advance, thus a different strategy is needed in their management. This can be seen as a basis for the precautionary principle, which for such systems means to maintain—even to work to enhance—system resilience.

Let me briefly return to physics. In 1999 appeared the English translation of a short book by Klein and Lachièze-Ray, *The Quest for Unity*.[4] In his review in *Nature* (Ellis 1999) George Ellis notes that currently "there is tremendous excitement among particle physicists, who are trying feverishly to develop a unified theory of all interactions and particles, and believe they are getting closer and closer to that goal. But the reality is that physics is becoming more and more fragmented." In a recent essay in *Nature* (Ellis 2005) *Physics, complexity and causality*, he extends his analysis. He notes that in a typical hierarchy of complexity, each level is linked to levels below, starting from basic physics, moving through chemistry and biology, to health and the behavioral sciences. The crucial word is "linked". Strict reductionism asserts that causation always works upwards, thus "link" has a causal meaning in the sense that each level is explained in terms of levels below—physics is all there is. But the world is more complicated than the physicists like to believe. They agree that in a complex system equations are written separately on each level, but such equations should—at least in principle—be reducible to more basic levels, just as thermodynamics is reducible to statistical physics. This may be fine when equations are linear, but in the non-linear cases we see new phenomena emerge, which are not reducible to component parts—ecological systems and the cognitive sciences offer many examples. The fundamental point here is that in complex systems we may have both upwards and downwards causality. Computational quantum chemistry may still have some success in explaining the properties of crystals from the behavior of individual atoms. Few believe that this is true in biology, cognition and behavior; see the discussion of the cognitive neuroscience in G. Marcus, *The Birth of the Mind* (2004). It should not be necessary to belabor the point that a sensitivity to the issues discussed in this section ought to form part of the background knowledge of every young scholar. But it requires that the same scholar be familiar with the necessary analytical tools—the mathematical sciences including statistics and computing.

[4]Klein, E. and Lachièze-Ray, M. *The Quest for Unity.*

4.3 The Third Way

Let me start with a story from *Nature* (Schilling 2001). The setting was a meeting on galaxy formation at Leiden in May 2000 and the speaker was just exhibiting an image of a cluster of galaxies when a member of the audience shouted out: "Is that an observation or a simulation?" The question was appropriate. As *Nature* noted, cosmologists have already created entire universes within computers, and they are now busy investigating such phenomena as asteroid collisions and supernovae—not in nature—but in their virtual universes. A new type of science seems to have been born. In addition to theory combined with observation and experiments, we now see computations and simulation as sources for insights and predictions about nature, society, and man. For a late update from cosmology see the article *Simulations of the formation, evolution and clustering of galaxies and quasars* (Springel et al. 2005).

To understand this trend it is necessary to take a step back to mathematics, or rather to the combined art and science of mathematical modeling. Mathematical modeling involves three main stages, first an analysis of a given scientific or technological problem, next the equations, i.e. the setting-up and analysis of a mathematical model for the system under study, and finally the development of efficient algorithms which can be used for predictions in the scientific context, and to constructions in the industrial one. In only simple cases can we solve the equations explicitly and compute with certainty the future development of the system. In most cases we have to resort to simulations. And with the rapid development of computer technology, including sophisticated tools for graphical representation, the range and usefulness of modeling is vast—we may mention natural language technology, financial markets, ecological systems and climate, the spread of infectious diseases, complex chemical reactions, the creation of the universe, construction of large structures, optimal extraction of oil—the list can be indefinitely extended. Let me stress one point, mathematical modeling is more than computers and codes and software, these are the final steps in the implementation of algorithms, and algorithms presuppose a mathematical analysis.

There are many reasons why the third way—modeling and simulations—is a useful extension of traditional scientific methodology. A good model gives you access to a wider range of initial data and parameter values than can be obtained by direct observations and experiments, and it allows you to explore time developments of the system which are not realizable in the real world; see the reference to computational cosmology above. But the main advantage appears in the study of non-linear dynamical systems, which include almost all systems in biology and the environmental sciences. The linear equations and deterministic laws that characterized classical physics, entailed exact predictions—the future is computable. With non-linear equations there is no longer a guarantee for long-term prediction. But we can do simulations and explore possible future developments. So instead of predicting with certainty we can prepare for the possible, even for adverse effects. This is what we are doing in environmental modeling, and this was precisely the insight behind the concept of *resilience* in ecosystems.

A word of caution is necessary. At a very early stage in modeling and computations one meets with the so-called GIGO principle (garbage in, garbage out) principle. Successful simulations depend on believable equations, which means that we have an adequate and sufficiently complete description of the system. Cosmologists seem to believe in their equations and hence assume that computations in a virtual universe give valid results for the one we inhabit. Others seem equally convinced; I cannot resist a quote from a recent review article, *A perspective on surfaces and interfaces* (Allara 2005): "In a field that has mainly emerged from empirical observations and experiments, it is now not unusual to see that well-executed theory and simulation can be accepted as more useful than experiments." Greater doubts may be entertained when we come to the social and behavioral sciences. Mass phenomena in the financial market can be successfully modeled by stochastic differential equations; we may not be so convinced about the rest of economics, despite recent Nobel prizes to game theorists. What is beyond doubt is that a thorough understanding of the possibilities, but also of the limits of the third way, is necessary for every young scholar today.

4.4 And the Next Century Is ...

Environment, biology and health were the theme of 18 of the 25 top questions asked by *Science*. We can argue whether the next century will primarily be known as the century of the environment, as discussed in the AAAS Presidential Address of J. Lubchenco, *Entering the Century of the Environment: A New Social Contract for Science* (Lubchenco 1998) or be known as the century of the biomedical sciences. I want to answer yes to both, but shall focus on the biological sciences in the following remarks.

An editorial article in *Nature*[5] asks a somewhat unexpected question: *Can biological phenomena be understood by humans?* The question stems from the Nobel physics laureate Robert Laughlin who had been invited to a conference earlier in 2000 to discuss quantitative challenges in the post-genome-sequence era. And seen from a physicist's point of view the question was not unreasonable. Let me tell a story to argue the point. In the history of astronomy we can clearly distinguish three stages identified with the names of Tycho Brahe, Johannes Kepler, and Isaac Newton. Tycho Brahe was the eminent observer of his age. He built the state-of-art instruments and he recorded over a long period of time precise data about the heavenly bodies. But he excelled only in data collection, and his attempts towards theoretical explanations did not go very far. Brahe's data were inherited by Kepler. He saw fixed patterns in the "sea of data" and succeeded in the task of describing the laws of planetary motion. But something was missing, the dynamical understanding of the observed patterns. This was the final step provided by Newton; he wrote down the "correct" dynamical equations, which explained the elliptical form of the Kepler orbits, which in turn had emerged out of a study of the data collected by Brahe.

[5]Can biological phenomena be understood by humans?—Editorial (2000) *Nature*, **403**, 345.

The analogy with the current situation in biology should be immediate and was, indeed, the theme of one of the questions posed by *Science*: How will big pictures emerge from a sea of biological data? Let me also note that the same story with the same moral has also been used by May (2004), and there have been a number of responses. The developing field of bioinformatics where much of this work is simple pattern analysis and classification, and the information that this generates has its important uses, is a good example. Even if we do not understand the "laws" that creates the patterns, we can use this information to tell a number of plausible stories about causes and effects, and we can even recommend new medical treatments and design novel drugs. This is no small accomplishment as the following quote from *The Economist Technology Quarterly*[6] shows: "In life-sciences establishments around the world, the laboratory rat is giving way to the computer mouse—as computing joins forces with biology to create a bioinformatics market that is expected to be worth nearly $40 billion within three years."

Beyond pattern analysis we are witnessing a major effort to create the equivalent of the Newtonian third stage. The catch phrase is *system biology*, see two recent reviews by Kitano (2002) and Pennisi (2003). The aim of these efforts is well explained in the introductory section of one of the papers in the review by: "To understand biology at the system level, we must examine the structure and dynamics of cellular and organismal function, rather than the characteristics of isolated parts of cell or organism. Properties of systems, such as robustness, emerge as central issues, and understanding these properties may have an impact on the future of medicine. However, many breakthroughs in experimental devices, advanced software, and analytical methods are required before the achievements of system biology can live up to their much-touted potential." Note that resilience discussed above, is a robustness property. For a specific example of current system biology, I recommend the article by Noble (2002).

System biology is, in all meanings of the words, a complex science and requires the participation of researchers from many disciplines. The challenge is to create research communities with shared knowledge, mutual understanding of the potential and limitations of a vast array of tools and methods, and with common culture and insights that transcends the original disciplinary boundaries. This is not easy, and there are pitfalls. As a mathematician I am particularly sensitive to problems of applying mathematics to other fields. Let me mention one example, the book *Mathematics in Population Biology* reviewed by Cushing (2005). The first half of the 20th century has been characterized as "the Golden Age of mathematical ecology and population dynamics", which gave us the basic predator-prey equations of Volterra and the so-called competition equations of Lotka-Volterra. However, despite this initial successful meeting of minds across disciplinary boundaries, traditional cultures came to dominate. When viewed as mathematics the topic became a minor activity within that discipline, and when seen as biology it became increasingly less relevant. The later results of "the meeting of minds" were mathematics that meant little to the ecologists.

---

[6]*The Economist Technology Quarterly* (14 Dec 2002), 30.

We have made our point, if system biology is the biology of the next century, then the internal research cultures traditional disciplines will have to change; this in turn will have a profound impact on how we in the future organize basic research training and plan the early careers of young scholars.

4.5 The Formative Years of Scholars

The American graduate school was one of the great successes of the post-World War II years. It demanded broad knowledge of facts and a mastery of many skills. It was well financed and was able to draw upon the talents of a whole world. I have personal experiences from Berkeley and Stanford, as a graduate student, as a young postdoc and later as a visiting scientist. And I did what many scientists with the same experiences also tried to do, work toward a similar structure of research training in your home country. It is fair to say that we had some success. But time does not stand still, and as I have tried to argue above, new and internal developments in science requires that we rethink the formative years of scholars. There are also external developments which may force changes to research training. The market place has moved closer, and, of equal importance, society requires that the traditional freedom of research be matched with an equally strong commitment to value issues and responsibility in the conduct of science; see the AAAS address by Lubchenco referred to above and the *Declaration on Science and the Use of Scientific Knowledge* from the World Conference on Science in Budapest, 1999.[7]

Following-up on the discussion in the last section I shall restrict myself to some remarks on the biological sciences and on internal developments in those sciences. Education on all levels has been extensively discussed within the profession. Three recent contributions from *Nature* (Dill 1999) and *Science* (Spengler 2000; Gross 2000). The message is unambiguous and well expressed in an editorial in *Nature Medicine*[8]: "A dominant theme emerging from our respondents was the growing influence of physics, mathematics and engineering on biology, and in particular how these disciplines will help biologists interpret their huge and growing data sets". But as the editorial also notes, it is not sufficient for the biomedical researcher to call upon the service of external hands, like technicians of yesterday. "For the convergence of physics and biology to really take hold, grow and bear fruit for the biomedical and ultimately medical community, biologists must become more familiar and comfortable with quantitative skills and grow more mature in their approach to mathematics and physics". For some experiences and specific advice on mathematics and biology, see *Mathematics Education for 21st-Century Biologists* (Bailek and Botstein 2004). We should also listen to the advice of F. C. Kafatos, former leader of the European Molecular Biology Laboratory (EMBL) and newly

[7](1999) *Declaration on Science and the Use of Scientific Knowledge* from the World Conference on Science in Budapest.

[8]*Nature Medicine* (2000), **6**, 113.

elected Head of the European Research Council, as expressed in an editorial in *Science* (Kafatos 1998): "The second challenge to European biology is structural inflexibility. This includes organizational structure and disciplinary conservatism in academia; excessive reliance on pyramidal power structures, when scientists need independence in their creative 20s and 30s; neglect of advanced postdoctoral training through overemphasis on lifetime employment; and top-down reflexes rather than trust in investigator initiative coupled with critical peer review. Despite the evident advantages of centers of excellence with critical mass, some countries succumb to the temptation to 'spread the rain evenly'." This was written in 1998 and much may have changed, but I expect no dissenting voice when I just take note of the fact that even more remains to be done.

My task has been to report on some recent experiences from the natural sciences, not to present a new program for research training and postdoctoral careers. But certain points stand out as challenges for the future reform work:

i. *Research training*: We see the necessity of a broader knowledge base, spanning many different disciplines. In an additive model of research training this will lengthen the period of study to almost infinity. We need an integrative model—but how should this be done?
ii. *Tools of many trades*: We have noted the vastly increasing repertoire of skills needed for the successful future research career. But tools need to be sharp; the "least common factor" model will never promote excellence—a "dollop of math" is no cure.
iii. *Critical mass*: Critical mass is necessary for building a shared culture across traditional disciplinary boundaries. This does not mean the denial of individual genius; the challenge is to create the right breeding ground for insight and creativity, and in this way also achieve the goals set out in the two previous points. Here we can draw upon many experiences from many different disciplines; we need to pursue a policy of "best examples" across traditional disciplinary boundaries in creating good research environments.

References

Allara, D. L. (2005). A perspective on surfaces and interfaces. *Nature, 437,* 638–639.

Bailek, W., & Botstein, D. (2004). Introductory science and mathematics for 21st century biologists. *Science, 303,* 788–790.

Cushing, J. M. (2005). Review of mathematics in population biology. *Bulletin of the AMS, 42,* 501–504.

Dill, K. A. (1999). Strengthening biomedicine's roots. *Nature, 400,* 309–310.

Ellis, G. (1999). No truth on grand unification. *Nature, 401,* 527–529.

Ellis, G. (2005). Physics, complexity and causality. *Nature, 435,* 743.

Gross, L. J. (2000). Education for a biocomplex future. *Science, 288,* 807.

Kitano, H. (2002). System biology. *Science, 295,* 1662–1664.

Kafatos, F. C. (1998). Challenges for European biology. *Science, 280,* 1327.

Le Bris, C., & Lions, P.-L. (2005). From atoms to crystals: A mathematical journey. *Bulletin of the AMS, 42,* 291–363.
Lubchenco, J. (1998). Entering the century of the environment: A new social contract for science. *Science, 279,* 491–497.
Marcus, G. (2004). *The birth of the mind*. New York: Basic Books.
May, R. M. (2004). Uses and abuses of mathematics in biology. *Science, 303,* 790–793.
Noble, D. (2002). Modeling the heart—from genes to cells to the whole organ. *Science, 295,* 1678–1682.
Pennisi, E. (2003). Systems biology: Tracing life's circuitry. *Science, 302,* 1646–1649.
Scheffer, M., Carpenter, S., Foley, J. A., Folke, C., & Walker, B. (2001). Catastrophic shifts in ecosystems. *Nature, 413,* 591–596.
Schilling, G. (2001). Exploring a virtual universe. *Nature, 414,* 12–14.
Spengler, S. J. (2000). Bioinformatics in the information age. *Science, 287,* 1221–1223.
Springel, G., et al. (2005). Simulations of the formation, evolution and clustering of galaxies and quasars. *Nature, 435,* 629–636.

Chapter 5
Remarks on the Science and Technology of Language

Abstract Language and logic are inseparably intertwined in the European intellectual tradition. It has not always been an easy relationship. There have always been issues of substance whether the two were friends or foes. Today computation is part of the equation, and we are seeing the development of a new discipline of computational linguistics. But meaning in language is more than combinatorial properties of signs. We note some limitations of the current technology and put forward some observations on the broader topic of grammar and logic.

Keywords Language and logic · Structural linguistics—Bloomfield—Harris
Chomsky—the revolution · Second revolution—Curry—Montague
Symbols—meaning

In most cases there is a reasonable correspondence between a subject matter and a science. Take physics as an example; there is little or no uncertainty about what the subject is and who are to be counted among the professionals. In almost all universities there is a central department of physics; nationally and internationally there are well defined professional associations, up to the level of the European Physical Society and the International Union of Pure and Applied Physics. There may be a bit of uncertainty about the exact line of demarcation between physics and some neighboring disciplines—such as chemistry among the natural sciences, and electrical engineering among the technical sciences—but a combination of subject matter, cultural background and current sociology of the profession uniquely determines where one belongs, whether you—to take one example with unexpectedly strong feelings involved—are a mathematical physicist or a plain mathematician working on mathematical problems in physics. No one would mistake the great Göttingen mathematician David Hilbert for a physicist even if he, independently and at the same time as Albert Einstein, arrived at the correct equation for the general theory of relativity.

The study of history offers another example. There is a core of the field focusing on political history, surrounded by spin-off specialities such as history of technology and even the history of science. Here also a combination of subject matter, cultural background and educational and professional behavior determines where one belongs. One would never mistake a political scientists for a historian. Of course,

J. E. Fenstad, *Structures and Algorithms*, Logic, Argumentation & Reasoning 15,
https://doi.org/10.1007/978-3-319-72974-9_5

in everyday life you may see a mathematician and a physicist, or a historian and a political scientist work peacefully side by side. But they know who they are and where home is; if challenged, they would at most describe their current activity as some 'problem-driven multidisciplinary project', if they are at all sensitive to what is politically correct. We may safely assert that there are few forces that are as strong as the gravitational pull of professional pride and property.

There may, however, be a counterexample. I have never been able to decide which science has the birthright to the study of human language and speech. Most professors of linguistics would claim this right for their own speciality; indeed, in most universities we will today find a department of linguistics—and is not linguistics the science of language and speech? Going beyond such superficial rhetoric we recognize that the case of language and science is remarkable more complex. In the European university tradition of the 19th century there were no separate departments of general linguistics. Language and literature were always the organizational unit, and one such unit existed for each language or language family. It was the need of the *gymnasium* which determined the structure of university departments in the humanities. In the emerging national states of Europe there was a special emphasis on the national languages and literature, and this had implications for education on all levels. The classics did not disappear; there were separate departments for the classics as well as for the major modern foreign languages. For the rest there were various ad hoc arrangements; e.g. in the the Nordic countries there would typically be a single unit for all the slavonic languages. The subject matter was dominated by the humanistic heritage of philology, with the usual emphasis on textual interpretation. After the 'discovery' of Sanskrit, the linguistic curriculum was enriched by historical and comparative studies. Such was—in simplified terms—the 19th century picture; for a comprehensive survey of the period I recommend H. Pedersen's *Linguistic Science in the Nineteenth Century*. But language, in particular seen as a cultural phenomenon, soon came into focus in other disciplines. There is an obvious link to anthropology; indeed, general linguistics, as distinct from philology and comparative studies, sometimes found its first home in schools or departments of anthropology; this was not uncommon in the Anglo-American tradition. The reason is perfectly clear, language as a general cultural phenomenon cannot be bound by a tradition with roots in the study of Latin and Greek.

With the emerging social sciences we also saw other interests staking claims on language as a field of study; psycho-linguistics focused on language and the individual; socio-linguistics on language and society. It alway was, and still is, unclear where the new hyphenated disciplines should belong. One should be careful in making bold generalizations, but my impression is that language as a cognitive phenomenon was more central to psychology than language as a social phenomenon to sociology.

This is not the end of the story. The study of speech was always an integral part of linguistics. But with the advent of 'modern' telecommunications, i.e. radio and telephone, the transmission of speech, seen as an important example of the technology of signal transmission, has become a recognized engineering speciality. Originally there were few links between the technology and the 'core' science of linguistics. Today telematics, seen as part of the expanding information technology, has progressed beyond the original signal transmission stage and faces new demands for more comprehensive technologies for language processing. One would assume that such technologies should rest on a broad science of language, and there is currently many activities devoted to a proof of this point. The matter, however, needs careful thought; I shall return to the issue at the end of this chapter. At the moment I will only note that speech recognition, which is a central component of the technology, is almost totally independent of current theoretical and computational linguistics. We do have effective systems for speech recognition, but the current technology is basically statistical in nature, making use of Bayesian transformations, hidden Markov-chains and frequency analyses of large text corpora—but it makes almost no use of grammar.

We have placed the study of language at the crossroad of various humanistic, social science and engineering disciplines. If we also want to see language as a brain-determined activity, we need to add several of the natural and medical sciences. It is, for many reasons, important to understand how the humanities and the sciences are interacting in modern society, but rather to reflect in general on this, I have chosen to concentrate on the specific case of language. This adds the richness of a specific context. I would also argue that whatever general points can be made on the relationship between humanities and sciences, can be made with equal force in our special case.

In the catalogue of contenders for the ownership of the science of language the reader may have noticed the omission of one discipline—logic. Language and logic have been inseparably intertwined in the European intellectual tradition. It has not always been an easy relationship, but there were always issues of substance whether they were friends or foes; some observations on the history of this relationship will be covered in the next section.

5.1 Some Observations on the History of Logic and Language

The starting point for this reading of the history is, of course, the Greek heritage, although the history of the systematic study of language is older and has more complex roots than the Greek experience. The systematic study of language was part of the general theoretical awakening in ancient Greece, with Plato and Aristotle as the central actors. Bur whereas the history of mathematics and the natural sciences has always been an integral part of our cultural history, the history of linguistics has remained a concern for the specialists. This is a bit curious seen against the background of the dominating role of grammar, logic and rhetoric as a part of the *trivium* of classical learning, the basis for higher education in Europe through the centuries.

After Aristotle there was a separation of ways in the study of grammar. One line of development was the philosophical or theoretical study developed by the Stoics and transmitted to the medieval grammarians through the rediscovery of Aristotle by

Boethius. The other line was the Alexandrian tradition represented by grammarians such as Dionysius Thrax and Apollonius Dyscolus. This was the so-called literary or descriptive way, which was particularly aimed towards the teaching of language and which built upon extensive examples drawn from the classical literary heritage. We can already at this point see a first example of a split in linguistic between a data-driven and a theoretical approach.

This account is not intended as a general history, but represents a personal view of the field. I shall therefore focus on what is necessary for later parts of this discussion. My next stopping point is at the beginning of the 14th century. Grammar had, in the early medieval period, been dominated by the descriptive line, focusing in particular on language teaching. Time was now ripe for a rebirth of a theoretical grammar. We cannot follow all the complexities of the historical development; the interested reader is invited to consult the edition of *Grammatica Speculativa of Thomas of Erfurt* by G. L. Bursill-Hall. For our purpose I note that a group of grammarians, the so-called *modistae*, with Thomas of Erfurt as a leading representative, managed to give a theoretical foundation to an existing descriptive practice. In doing so they built upon the rediscovered Aristotle, but went far beyond in creating a new theory. With the help of generous hindsight we can recognize in the theoretical analysis of the *modistae* both a grammatical surface structure and a semantical-logical deep structure. The grammar of the *modistae* is, beyond doubt, a high point in the scholastic edifice of learning. Today their theory is, for many, a closed world, embedded as it is in scholastic philosophy. But in my judgement Thomas of Erfurt is exemplary in his attempt to give a connected account of structure and meaning.

For the current observer of the linguistic scene it may look strange to stop at the strong medieval brew of Thomas of Erfurt. Why not, following current fashion, trace our 'revolution' in linguistics and language engineering to the clear cartesian foundation found in the Port-Royal grammar of circa 1680? I would have liked to agree; I have, with great enthusiasm, read the *Grammaire Générale et Raisonné* by the Port-Royal Jesuits Lancelot and Arnold. But one needs to show some care with the historical scholarship of the much celebrated book *Cartesian Linguistics* by N. Chomsky; see for example the critique in G. A. Padley *Grammatical Theory in Western Europe 1500–1700*. We may now with some confidence claim that it was the *modistae* and not the Port-Royal grammarians, who had the fresh and original insights, and that it was they who had a perspective which again today is productive in theoretical linguistics. But before this could be appreciated linguistics once more divorced itself from logic and turned back onto a descriptive road.

We next make a jump to the 1930s. The excessive historical and comparative studies of the 19th century had at that time been replaced by a structuralist approach, which was initiated by the great Swiss linguist Ferdinand Saussure. We shall take the well-known American linguist Leonard Bloomfield as our witness. As a structural linguist he occupied a central position in his discipline towards the end of the 1930s. Philosophically he subscribed both to empiricism and behavioralism. In an article *Linguistic Aspects of Science*, which he contributed to the *International Encyclopedia of Unified Science*, he gave an almost militant expression to his convictions. In his brief survey of the history of the subject he also starts with the Greeks, but his tone is severely critical, accusing them of serious mistakes due to their unscientific (read: metaphysical) definitions. Following Bloomfield, modern linguistics should trace its roots to the rediscovery of Sanskrit and to the ancient grammar of that language by the Indian grammarian Panini. We see in Panini, still according to Bloomfield, a descriptive—even 'scientific'—grammar, rid of philosophical assumptions. This kind of grammar set an example for the description of native Indian languages in America, and it provided a historical foundation for the data-driven and inductive approach of the structuralist school in American linguistics. Bloomfield's contribution to the *Encyclopedia* is fascinating reading. If we only could believe in him, the final victory was at hand. But read against the background of current insight into the cognitive sciences, his behavioristic foundation for linguistics is seen to be a remarkably primitive stimulus-response construction.

American structural linguistics saw itself as the culmination of linguistics as a science. The book that was intended to put the last stone in its proper place, was *Methods in Structural Linguistics* by Zellig Harris, that appeared in 1951. The methodology is sketched out in an introductory chapter. The procedure is strictly inductive and based on a comprehensive distributional occurrence analysis. The structure of the particular language under study must be read off from the inductive analysis. The theory of linguistics has de facto been reduced to an procedure for language description. What remains of meaning is at most an epiphenomenon, a pure case of stimulus-response behavior, which is objectively observable by the field linguist. There is no universal theory of grammar and meaning.

The brevity of this description is unfair to what Harris did. He was a great linguist with fundamental insights to his credit. Experimental design and descriptive procedures are an important part of any empirical science; read in this light his contributions are important. He also indicated the need for a mathematical study of the elements which emerged from the inductive analysis, and he pointed to the importance of mathematical transformations in the final structural description (e.g. an active-passive transformation). The seeds of a generative analysis are thus implicit in some of his theoretical considerations.

We now move forward to the year 1957. This year marks the beginning of Noam Chomsky's revolution of linguistic science. We should, however, note that there were some exceptions. The revolution was, in the beginning, an American phenomenon—there were European traditions which long remained outside the revolutionary circle. But the impact of Chomsky fully justifies the word 'revolution'.

The ground was well prepared. The philosophical foundation, i.e. the extreme behaviorism of Bloomfield, had started to crumble. Skepticism concerning the possibility of a pure inductive determination of structure was also more pointedly raised; some critics accused the structuralists of secretly looking at meaning and other mental phenomena in the determination of structure; they pointed especially to the suprasegmental analysis of intonation. On the other hand, one had seen the start of a formal analysis in the work of Harris, and modern developments in mathematical logic, in particular, in recursion theory and the theory of formal languages, provided the tools necessary to translate such linguistic insights into 'hard' science. The bits were assembled into a coherent picture by Noam Chomsky: first in a short monograph

Syntactic Structures from 1957; a few years later in a comprehensive and convincing manner in the book *Aspects of the Theory of Syntax*. This is not the place to give an introduction to Chomskian generative linguistics. We only note that within a couple of years there had occurred a revolution in linguistics. A historian of science would probably see some parallels between Chomsky's *Aspects* and the influence of its time and Thomas of Erfurt's *Grammatica Speculativa* and the influence of the *modistae*.

The analogy is only partial. Chomsky argued in principle for an equal status of a syntactical, a phonological and a semantical component, but in his actual model building it is the syntax and phonology that dominate. The reason may have to with the mathematics available at the time; this a point I shall return to later.

Here I wish to emphasize that the Chomskian revolution was not only an internal issue of content, it was equally a cultural and sociological revolution; the reader may wish to consult F. J. Newmeyer's book on *Linguistic Theory in America*. But, simultaneously with the Chomskian transformation of the linguistic landscape, there occurred another 'revolution'. It was hardly visible, and it took the linguistic profession almost 20 years to digest what had happened. In retrospect, the second revolution can make a stronger claim of being the 'true' successor to the revolution of the *modistae* of medieval times.

Modern logic is more than axioms and deductions. Semantics was always an integral part; in modern times we can draw a line from the German logician G. Frege via the Polish logician A. Tarski to the current theory. John Stuart Mill wrote in 1867 that 'the structure of every sentence is a lesson in logic', but the structure of logic seemed at first to be much too rigid to serve as a tool for linguistic analysis, except at a sufficiently deep logical level. The logical empiricists of the 1930s also held natural language in low regard, as the following programmatic pronouncement of C. W. Morris, one of the chief architects behind the *Encyclopedia*, bears witness: '... it has become clear to many persons today that man—including scientific man—must free himself from the web of words which he has spun and that language—including scientific language—is greatly in need of purification, simplification and systematization. The theory of signs is a useful instrument for such debabelonization'. Let us add that logical syntax and semantics in the style of Rudolf Carnap was to be the main tool in this task; everything that did not fit into this scheme should be demoted to behavioristic stimulus-response models. It is remarkable that Bloomfield was a co-contributor to the same unified program.

But mathematical model-building is more than the naive formalizations of the early empiricists. One of the first to apply the whole range of tools of modern logic to the study of natural languages was the German philosopher of science Hans Reichenbach, who included a chapter on 'conversational languages' in his textbook *Elements of Symbolic Logic* from 1948. The particular tool that Reichenbach used was a version of higher order logic extended by certain 'pragmatic' operators.

From about the same time we find a contribution by Haskell B. Curry, an American logician and one of the last to receive his doctorate in Göttingen under the supervision of Hilbert. Curry lectured on his results in 1948, but a paper was published only in 1961. In this work Curry presented an analysis of traditional grammar using combinatory logic. The analysis is related to, but independent of, the 'categorial analysis' presented in the 1930s by the Polish logicians S. Lesniewski and K. Ajdukiewicz. Curry's main advance compared to the Polish logicians is that his combinatorial logic has a naturally associated semantics in terms of a higher order functional structure.

This was, in retrospect, the second 'revolution'. But as remarked above it was hardly visible. Curry remained unpublished; Reichenbach's book appeared a few years before Harris' Methods in Structural Linguistics, and must for the contemporary linguist have represented a step back to the *a priorism* of logic and metaphysics. But linguists ought to have read their Reichenbach with more attention. They would then have noticed that he gave the first satisfactory analysis of meaning of natural languages, using higher order logic. He broke through the narrow framework of standard predicate logic, which long had been recognized as inadequate for linguistic analysis. The pragmatic operators he introduced, were a first approximation to later developments in intentional logic. The analysis of time which he gave, is still the foundation for present day linguistic theories. The main complaint against Reichenbach, but not to the same degree against Curry, is the lack of an explicit syntactic component. But even here the tools were available at the time; I am thinking of the categorical grammar of the Polish logicians.

Few people have read Lesniewski's paper from 1929 on *Grundzüge eines neuen Systems des Grundlagen der Mathematik*, his ideas were made known through the work of Ajdukiewicz *Die Syntaktiche Konnexität* from 1936. This line of development was carried forward by the Israeli philosopher of science Y. Bar-Hillel, who wanted to use categorial grammar in machine translations around 1950: more on this below.

The second revolution remained at the time a possibility. The necessary elements were at hand, but no one saw how to fit the pieces together in a picture that would convince the linguistic community. It could have been Bar-Hillel, the logician who knew his Reichenbach and who wanted to apply categorial grammar to machine translation. But Bar-Hillel, after the early failure of machine translation, became towards the end of the 1950s an enthusiastic supporter of Chomsky; thus categorial grammar, combinatory logic and higher order structures disappeared from center stage. Posterity may regret what happened. We may feel that the lack of insight was remarkable, in particular, since all pieces had the same ancestry, which was the theory of types of the *Principia Mathematica* of Russell and Whitehead from 1910. It took 20 years before the pieces were finally put together. This was achieved in a series of works by Richard Montague from the mid-1960s; see his collected essays *Formal Philosophy*. Why did some fail and others succeed? This is an issue which involves both mathematics and the sociology of the profession.

5.2 Linguistics and the Mathematical Sciences

We shall look in some details at a few modern episodes in the sometimes turbulent relationship between linguistics and the mathematical sciences. For present purposes our starting point will be the de facto separation of ways in the mid-1930s.

In linguistics, American structuralism's was a leading paradigm. Linguistics was reconstructed as an empirical science, proceeding in an inductive way in its task of extracting structure from data. Meaning was an added epiphenomenon of an objectively observable stimulus-response behavior. The other partner of our story, philosophy, has always been more confused, but one leading paradigm of the 1930s was the school of logical empiricism. A leading spokesman, first in Europe, later in the US, was Rudolf Carnap, who was also one of the main forces behind the *International Encyclopedia of Unified Science*. Carnap would obviously recognize the need for an empirical study of language as a cultural phenomenon, but he would no doubt subscribe to the pronouncement of Morris that 'language—is greatly in need of purification, simplification and systematization'. Carnap's own contribution to the needed 'debabelization' was the severely formalistic tome on *The Logical Syntax of Language*. Hans Reichenbach was also one of the leading scientific philosophers, well-known for his contributions to the foundation of probability, to the study of space-time and the foundation of quantum mechanics. Also highly respected is his study on the irreversibility of time. We have praised his contribution to linguistics, but to his contemporaries the chapter on 'conversational languages' was only a minor, perhaps, even curious episode in a rich scientific career.

Not every linguist marched under the banner of structuralism. Let me note one exception, the Danish linguist Otto Jespersen. He wrote in 1920s a book, *The Philosophy of Grammar*, which was at odds with the structural thesis and much more in line with traditional grammar. The book has been reprinted a number of times and is still read today. In 1938 he published a short volume on *Analytical Syntax*, in which he tried to give a formal expression to his view. But that book—even if it has been recently reprinted, since Jespersen is a famous linguist—is curiously barren in content. My explanation is simple; Jespersen did not know the kind of mathematics necessary for his task. He belonged to a long and distinguished tradition of Danish linguistics, which was dominated by the historical and comparative themes of the 19th century, but which gradually changed to the structuralist point of view. Jespersen had his independence of mind, but he was bound by his educational heritage. In no way was the acquisition of a broad mathematical culture part of the upbringing of a European linguist.

It is instructive to contrast this episode with the case of Haskell B. Curry. We met Curry above as one of the silent heroes of the second revolution. His starting point was his familiarity with traditional grammar, and he possessed the mathematical tools necessary to give a formal expression to his insight into grammatical structure. To be slightly technical, traditional grammar is cast in a subject-predicate form; in fact, the elementary grammatical operations can be seen as examples of a function-application formalism. Combinatory logic, which was the speciality of Curry, is an applicative system particularly well suited for the formalization of traditional grammatical operations. This is a feature which it shares with the categorial grammar of Ajdukiewicz. But combinatory logic adds more. It has a natural associated meaning theory in the form of a higher order functional structure. Thus, Curry was able to provide a combined syntactical-semantical analysis of a fragment of English. Curry was no linguist, but his familiarity with traditional grammar was sufficient to produce a significant scientific insight. Curry had little impact on the linguistic community; he was quite isolated in his search for linguistic structure, neither linguists nor logicians paid much attention.

The state of Pennsylvania in the US was Curry's home territory. A few years later the same geographic location was the scene of the 'visible' revolution in linguistics. Noam Chomsky was, in a somewhat unusual sense, the student of Z. Harris at the University of Pennsylvania; see the introductory interview in the collection *A Chomsky Reader*. I do not know the exact influence of Harris on Chomsky, but as we noted above there are mathematical seeds in Harris' *Methods in Structural Linguistics* which, in a full-grown way, are present in Chomsky's *Syntactic Structures*. Similarly, I do not know if there were any serious contacts between Curry and Harris. Curry's paper was presented in a seminar for linguists. If it had provoked any response, Harris would have known. I find it fascinating to see the celebrated 'structural linguist' Harris interpolated in this curious way between the 'logician' Curry and the 'generative linguist' Chomsky in the late 1940s—early 1950s in the state of Pennsylvania.

Chomsky succeeded, partly because he was a master of the mathematics he used; indeed, he made several important contributions to formal language theory. But part of the success was also sociological. The linguist community was ripe for the elevation of their field of study to one of the 'exact' sciences; see the account in Newmeyer's book. Let me, however, add one slightly critical observation on this success story. Chomsky was, as one always is, constrained by the mathematics he knew. Meaning was always a recognized part of the Chomskian paradigm, but the mathematics he mastered, forced an almost exclusive attention to syntactical matters. Remembering Chomsky's attempt to link his linguistics theories to the Port-Royal tradition, one conjectures that he would have liked to use a full-blown mathematical development of meaning. But Tarskian semantics was still in its infancy around 1950, even Reichenbach had to resort to some ad hoc trickery to give an account of meaning.

Richard Montague was a student of Tarski in Berkeley in the late 1950s. He worked independently, but seen from our vantage point his contribution consisted mainly of making a coherent pattern out of the elements found in Ajdukiewicz, Curry and Reichenbach. I particularly recommend his paper on the *Proper Treatment of Quantifiers in English* as a substantial sample of what he did. Montague succeeded where Curry failed, and we can ask why—the insight and power of their respective mathematical analyses are comparable; they both could have succeeded. But Montague had an interpreter, Curry stood alone. Barbara Hall Partee is a leading American linguist. She was professor in the Linguistic Department of the University of California at Los Angeles at the same time as Montague was a professor of philosophy there. Whereas Montague saw his work as an exercise in formal philosophy, Partee saw the linguistic significance of his work. Through a series of studies she made his work accessible to the linguistic community. She also made important new contributions, highlighting the linguistic relevance of the formal apparatus. With her stamp of approval linguists had to take notice.

There is one more episode to tell; this time European, and it involves both a linguist and a mathematician. The linguist is L. Tesnièr, a French scholar working outside the main intellectual centers. Whereas the structure of Chomskian syntax was rather Aristotelian in form, Tesnièr attempted to find a pure relational form to the syntax, putting the verb phrase into a central position; see his book *Elements de Syntaxe Structurale* from 1958; in this respect he is in the line of Curry, Reichenbach and Montague. The syntactical studies of Tesnièr has inspired other European linguists, particularly in Germany; we mention the *Dependenztheorie* as one important example. Tesnièr's work also served as an inspiration for some interesting geometric reflections on grammar by the French mathematician René Thom. In two papers from around 1970, *Topologie et Linguistique* and *Langage et Catastrophes: Elements pour une Sémantique Topologique*, he used the syntactic analysis developed by Tesnièr as his linguistic background. The central role of the verb phrase in Tesnièr's approach is in Thom's approach translated into a classification of singularities in a suitably derived energy surface. This may sound odd, but is a 'correct' step toward a comprehensive theory of meaning in natural languages. There is no room here for a detailed discussion of grammar and geometry; the story is continued in Chap. 8. Thom's work did not have the influence on theoretical linguistics that it merits. There seems to have been an incompatibility of minds. Thom severely criticized the generative approach as being totally inadequate as a theory of linguistic meaning. Linguists, unfamiliar with the mathematics, saw little relationship between Thom's 'speculations' and their science. Do we once more see a weakness of the European university tradition, which has been dominated by a strong sense of professional purity, which in turn has tended to encourage a certain 'inwardness' in the education of young scholars? It seems that the American Graduate School at its best was remarkably effective in creating a broader intellectual community. This is not a question of a fashionable multi-disciplinary approach, but the plain task of learning the tools of many trades. Barbara Hall Partee had a sound training in mathematics in the graduate school of MIT.

5.3 Adding Computer Science to Logic and Linguistics

We have so far concentrated on the relationship between grammar and logic, i.e. on the theoretical understanding of how structure and meaning are connected. The new computer sciences have added new possibilities; suddenly there was a promise of 'true' engineering applications. 'Applied linguistics' existed before—there were some applications, for example, to speech disorder, to language teaching etc.—but applications were never a major concern of the community. Computers changed that.

The impetus came from an outsider. In 1949 Warren Weaver of the Rockefeller Foundation wrote a famous memorandum on the possibility of automatic translation. His first approach was to look at translation as a decoding task, taking his cue from the successful work on cryptography during the war. Add to this the possibilities of efficient large-scale symbol manipulation provided by the emerging computer technology and you have a promising research agenda. It was soon realized that machine translation could be a useful technology in cold war efforts, and suddenly money started to pour into an activity, which—to be honest—was never more than a promise.

One of the early actors in the translation enterprise was Y. Bar-Hillel, whom we mentioned earlier. It was Bar-Hillel's idea that categorical grammar was a good instrument for the decoding task, but the early hopes soon came to an end—the complexity of language structure goes beyond what can be captured by categorical rules. It is fascinating to follow the early history of machine translation through a series of reports written by Bar-Hillel, tracing the development from the optimistic visions of his first report on *The State of Machine Translation* from 1951 to the pessimistic appendix *Demonstration of the Non-Feasibility of Fully Automatic High Quality Translation* from 1960. The pessimism of 1960 almost killed machine translation, although some smaller communities survived. Bar-Hillel himself left the field and became a strong supporter of the Chomskian revolution of the late 1950s. One key innovations of *Syntactic Structures* was the concept of transformation. Adding transformations, such as the active-passive transformation, added power to Chomsky's analysis, but one should note that transformations are computationally rather intractable; indeed, there was for a long time, no computational praxis as part of the Chomskian enterprise.

A new lease on life for machine translation came with the European Commission's large-scale attack on the problem through the EUROTRA project of the 1980s. There were also some smaller projects in the US and Japan, but this is not the place for a comprehensive history of machine translation. It is easy to understand why the Commission should be interested in the development of systems for automated translation, and it is equally easy to understand the willingness of linguists, both theoretical and computational, to participate. But once more one was chasing a vision without the proper foundation in a technology—EUROTRA was never a success.

However, machine translation with more limited aims has survived as a useful technology. Text and the manipulation of text are an integral part of computer technology, and we have witnessed, sometimes in an ad hoc fashion, the emergence of a new trade, computational linguistics. Applications of this trade did not always proceed according to the book of linguistic theory; for limited task a mastery of the technology and skill in programming were more important than adherence to the fine points of theory. As remarked earlier, current systems for speech recognition make almost no use of linguistic theory. This should not come as a surprise, there is no difference in the relationship between linguistic theory and related engineering practices than we see otherwise in the relationship between a basic science and engineering applications. The interplay between science and technology is complex and not always well understood; there is seldom a straight road from basic research to successful technologies in the market place. We have today seen many useful applications of linguistic technology in the information sciences; but it is important to keep in mind that language technology, as any other technology, must live by its successes in the market place. There is a sound warning in an interview with G. Metakides, Director of the Information Technologies Program in connection with the

4th Framework Program of the European Union, 'we do not want to be responsible for producing technological white elephants. There are too many cases in the past of successful technological projects ... The R&D people got money from the outside; management was happy—but nobody used the results'.

We have at several places above noted a tension between data and theory in the study of language. Already the ancients recognized the tension between data-driven practice and theoretical speculations. We observed that the correct theory according to Harris was conceived to be almost entirely data-driven; inductive generalizations based on large text corpora was the only permissible methodology. Montague grammar is at the other end of the spectrum as an example of pure theory; a formula in higher order intentional logic, where truth is truth in all possible worlds, is not meant as a starting point for further computations. The opposition between theory and data is, of course, a false opposition; the science of language, as well as any other empirical science, needs both. We have commented above on the application of linguistic technology to the information sciences. There is an equally important converse, the application of information technology to the study and management of data in linguistics; the support of such activities is not primarily a responsibility of the market place.

Most useful techniques in syntactical and morphological analysis proceed at the level of symbol manipulation, such as sorting, ordering and comparing symbols, and listing of symbols. This is equally true for the application of language technology to the information sciences and for the application of information technologies to the study of language. But symbols carry meaning, and viable systems for machine translation and the analysis of dialogue require that we master a technology of 'computational semantics', which needs to be based on the science of the second revolution. In fact, this is what later community programs in language engineering seem to call for: 'RTD will concentrate on spoken and written language understanding, as opposed to such physical aspects as signal processing and character recognition' to quote from the Telematic's Work Plan of one of these EU Framework Programs. If you have learned the intended lesson from this reading of history, you will appreciate that this is no easy task, and that it still is—except in those limited context where meaning is reducible to a database listing—beyond todays engineering capabilities; for further discussion of these themes see Chap. 7.

5.4 Symbols Carry Meaning

Language never ceases to fascinate. Many sciences have the data of language as an important empirical foundation, but no science has an exclusive claim; this was our starting point. Others may prefer different perspectives in the search for explanation and insight; I have concentrated on the complex interplay of logic, language and computations. Language, seen both as a science and a technology, has been the theme of this chapter; we have been able to tell some success stories, but also to note some less glorious episodes. In a one of his speeches M. Carpentier, the former Director

General for DGXIII of the European Commission, noted that 'just as at the end of the last century and the beginning of this, a new resource—energy—revolutionized both economic activity and individual and social behavior, so now, as we approach the end of the 20th century, information technology and telecommunications are generating a second industrial revolution. This revolution is based on information, the expression of human knowledge'. If this is so, language—both written and spoken—must be part of the new technology. Linguistic signs have a combinatorial structure, but they are also carriers of meaning. The technology today can cope with the combinatorial properties. But meaning is not reducible to symbol manipulations. Thomas of Erfurt, circa 1315, could have told us so. More is needed—a challenge we shall return in Chap. 8 on *formal semantics, geometry and mind*.

Advice on Literature

The discussion in this chapter has touched upon many topics. A complete set of references would be lengthy and probably not of interest to the general reader; some documentation of sources is, however, necessary and some advice on the literature may, perhaps, be of interest.

The medieval rebirth of theoretical grammar is well documented in:

Bursill-Hall, G. L. (1972). *Grammatica speculativa of Thomas of Erfurt*. London: Longman.

Padley, G. A. (1976). *Grammatical theory in Europe 1500–1700. The Latin tradition*. Cambridge: Cambridge University Press.

General historical background on linguistics and logic with exact references to works quoted in this essay can conveniently be found in:

Fenstad, J. E. (1978). Models for natural languages. In J. Hintikka et al. (Eds.), *Essays on mathematical and philosophical logic*. Dordrecht: Reidel.

Newmeyer, F. J. (1980). *Linguistics theory in America*. New York: Academic Press.

Robins, R. H. (1967). *A short history of linguistics*. London: Longman.

Current developments on the interaction of grammar, geometry and the cognitive sciences are surveyed in chapter 8 below.

Chapter 6
How Mathematics Is Rooted in Life

Abstract Mathematics is almost always an insider's affair. But sometimes things happen within the mathematical community that have a relevance, and perhaps also an interest, beyond the tribe itself. The *Grundlagenstreit* of the 1920s is such an example. In this review essay we tell this story with focus on the main actors involved, David Hilbert in Göttingen and L. E. J. Brouwer in Amsterdam. We shall see how fine points concerning the existence of mathematical objects, the question of the editorship of the Mathematische Annalen, and the attempts to resume normal scientific contacts between French and Germans scientists after the First World War led to an unusual bitter conflict within the tribe and beyond. But even if the effects of the fight were at the time negative, the long range outcome was positive. Hilbert's work on the foundation of mathematics is still a powerful influence on current research, and Brouwer's view on the constructive foundation of mathematics, which at the time inspired both Husserl and Wittgenstein, is today of increasing importance in the evolving science of logic and computing.

Keywords Brouwer—Hilbert—Grundlagenstreit · Intuitionism—constructivity Type theory—programming languages

The title of this chapter is the subtitle of a recent biography of the great Dutch mathematician, logician and philosopher Luitzen Egbertus Jan Brouwer, known to his friends and foes alike as Bertus, written by Dirk van Dalen. A history of mathematics is almost always an insider project with little relevance outside the tribe itself; this book is an exception. We shall explain why.

Mathematics in the 19th century was very explicit, algebra was basically symbolic calculations, a function had to be given by an explicit formula, and if you asserted that something existed, e.g. a root of an equation or the value of a function for given arguments, it meant that you could actually compute the answer. Cantor's *Allgemeine Mengenlehre* changed this. Algebra became the study of abstract structures, functions changed from formulas to sets of ordered pairs, and existence came to mean that non-existence was contradictory.

Cantor's set theory was for most mathematicians a welcome expansion of the mathematical universe. In particular, it was important for the early development of general topology. However, difficulties soon appeared. First in the more distant and

J. E. Fenstad, *Structures and Algorithms*, Logic, Argumentation & Reasoning 15,
https://doi.org/10.1007/978-3-319-72974-9_6

not too important parts of logic and mathematics, e.g. the Russell paradox of the set X of all sets x such that x is not an element of x itself: Is X an element of X?

But soon the uncertainty about the soundness of set-theoretical reasoning moved closer to the central parts of mathematics. Dedekind had in the 1880s presented his axiomatization of the set N of natural numbers: 1, 2, 3, ... And by stipulating that the set N is the least set that satisfies the axioms, he could "prove" that N is completely characterized by his axioms. But this means that N is defined in terms of itself, since N is one of the sets that satisfies the axioms. This goes against our intuition that N is stepwise constructed by starting with the number 1 and then adding 1 successively to generate the full number series.

Objects defined in terms of themselves are called impredicative. Some saw impredicativity as the villain of set theory, prominent among these is the great French mathematician Henri Poincaré. His range as a mathematician was formidable, here we take note of him, relevant for our later story, as one of the fathers of modern topology. His views inspired a French school of "semi-intuitionists", including leading mathematicians such as Borel, Baire and Lebesgue. They wanted to understand the nature of an arbitrary subset of the real continuum, not by implicit impredicative definitions, but by stepwise building up an extensive hierarchy of definable subsets, now known as the Borel hierarchy, as an approximation to the full family of subsets of the reals. The resulting theory, today known as descriptive set theory, is an important part of current mathematics, both in praxis and in foundational studies.

Paris and Göttingen were the two main centers of mathematics. This was recognized by all in the mathematical profession in the late 19th century. The two communities lived side by side in a fruitful collaboration and competition and hardly noticed the war of 1870–71. This changed with the First World War, when also mathematics became part of a larger political game. This will be part of our history, but first we return to more peaceful times. At the turn of the century Paris became semi-intuitionistic, Göttingen remained set-theoretic. Hilbert's *Grundlagen der Geometrie* was an extraordinary success, and the axiomatic method became the chosen Göttingen method of how mathematics was to be done. Zermelo made an important step forward by his axiomatization of set theory in 1904. Next step would be to prove the axioms *consistent*, which means that the set theoretic universe exists, and finally that they were *categorical*, which means that the set theoretic universe is uniquely described by Zermelo's axioms.

Hilbert's proof theory was gradually constructed in the early part of the 20th century to achieve these goals. We shall return to this story, but let us at this point pause and review the situation in the years before the great war. Cantor's Mengenlehre had made a great impact on mathematical praxis and understanding, but there were difficulties. Two ways out were suggested, Hilbert's axiomatization with proofs of consistency and categoricity, and Poincaré's limitation to predicative methods of proof. Both ways were in a sense conservative, the classical kernel of mathematics was to be preserved, perhaps with certain restrictions and refinements. Hermann Weyl was a star student of Hilbert in Göttingen. It is possible to view his 1910 paper *Über die Definitionen der mathematischen Grundbegriffe* as an attempt to build a bridge between Paris and Göttingen. Thus there could be a harmonious way forward. But mathematics never fails to surprise, there will always be a new young voice confounding the wisdom of the elders: enter Brouwer in 1907, the year of his thesis *Over de grondlagen der wiskunde*.

Brouwer had the confidence of youth and talent. Hilbert had at the world congress of mathematics in Paris in 1900 given his great lecture on *Mathematische Probleme*, intended as a guide to the future development of mathematics. The challenge was accepted by the community, and your reputation as a mathematician was assured if you solved or contributed significantly to the solution of one of these problems. Brouwer claimed to solve three of these problems in his thesis: problem one in Hilbert's list on the nature of the continuum, problem two in the list on the consistency of the arithmetical axioms, and problem five in the list on a certain question of topology connected with Lie groups. Let problems one and two rest for a moment; we turn to problem five.

What Brouwer did, was to solve a special, but significant case of the fifth problem. This was main-stream mathematics and was instantly recognized. And it led to further work in topology. In a short period from 1907 to 1913 Brouwer, continuing the early work by Poincaré, created to a large extent modern topology, the Brouwer fixed-point theorem being a noteworthy example. Brouwer had a powerful geometric intuition and made little use of algebraic tools, which later replaced his more direct geometric approach. But his results stand and are the foundation for modern work. Hilbert at once recognized the new star, and in 1914 Brouwer was invited to be an editor of the Mathematische Annalen, the leading journal of mathematics at the time, and he became a honorary member of the Göttingen mathematical community. But this was before Hilbert recognized the "subversive" views of Brouwer on the foundation of mathematics. We turn to questions one and two in Hilbert's list, the nature of the continuum and the consistency of the arithmetical axioms.

Brouwer's view of mathematics was deeply influenced by German idealistic philosophy, mainly Kant, as can be seen from his first major published work *Life, Art, and Mysticism* from 1905. In his thesis from 1907 he characterized mathematics as the free activity of the human mind; thus mathematical truth is neither to be secured by reference to some abstract platonic realm beyond space and time, nor by reference to a formal game of axiomatics and consistency proofs. Mathematics was a meaningful activity of human thinking, it was primarily rooted in the intuition of structure and patterns. Language, formalism and axioms were secondary and could never be seen as a proper foundation. According to Brouwer's views a mathematical proposition can be recognized as true only when the thinking subject has experienced its truth through a mental construction; in particular, an existential proposition can be true only if you have an appropriate mental construction of an object satisfying the proposition. This led to the rejection of the ancient Aristotelian *law of the excluded middle*, that for every proposition A, either A or not-A is true. Accepting this principle of classical logic means, in our case, that to prove the consistency of arithmetics, which is Hilbert's problem two, it is sufficient to prove that inconsistency is impossible. This was rejected by Brouwer, existence means construction, not the impossibility of non-existence. He also rejected Hilbert's problem one concerning the nature of the continuum as ill-posed and meaningless. The notion of a continuum as a completed and uncountable pointset was denied by Brouwer and the core part of his intuitionism consisted in developing methods to "construct" points on the real line and to build a "sound" mathematical praxis on this foundation. Through this work he became a pioneer in the development of constructive mathematics. We shall below return to constructivity and its impact on mathematics and computing today; here we continue with the *Grundlagenstreit* of the 1920s. This foundational *Streit* inside mathematics became, perhaps as a surprise to many university professors, part of the broader cultural and political conflicts of the turbulent years between the First World War and the coming of Hitler's Germany.

At first Hilbert did not pay much attention to the critical Brouwer. For him Brouwer was the new star of topology, to be welcomed as editor of the Mathematische Annalen and to be a valued associate member of the Göttingen community. In fact, Hilbert at one point wanted Brouwer as the successor to Felix Klein. But when his star student Hermann Weyl in 1921 announced in a paper on the new foundational crisis in mathematics his conversion to Brouwer's intuitionism—this is the revolution—the old master was not amused.

Hilbert had after his Paris lecture argued for an axiomatic approach to the foundations of mathematics. Existing mathematics should be codified in an all inclusive axiomatic system with a well-defined syntax and exact rules of proof. The challenge was to prove such a system consistent. What was added in the 1920s in response to the criticisms of Brouwer and Weyl, was a more profound analysis of what it means to prove a system consistent. To prove consistency of Mengenlehre by unrestricted set-theoretic methods would carry no conviction. Geometry was proved consistent by Hilbert by reduction to an arithmetical foundation. But how to prove the consistency of number theory itself? Hilbert's proposal was to restrict consistency proofs to finitary methods, with no use of quantifiers (such as the universal "for all" and the existential "there is") over infinite domains, such as the set of all natural numbers.

Hilbert presented his ideas in a number of lectures during the 1920s, famous is his lecture in Hamburg in 1927 on *Die Grundlagen der Mathematik*. Brouwer did not directly respond to the sequence of Hilbert lectures; he presented his ideas to wider audience in Berlin in 1927 and Vienna in 1928. The fight between Hilbert and Brouwer engaged the mathematical community. In a larger context it is possible to see the Grundlagenstreit inside mathematics as part of the larger struggle between the ancient regime, and the newer revolutionary ideas emerging after the First World War. Hilbert spoke for the status quo, Brouwer championed something new, which could be a revolution in how to do mathematics. We know that Wittgenstein was in Brouwer's audience in Vienna, and it is suggested that his return to philosophy was inspired by Brouwer. Wittgenstein went forward and never looked back. Hermann Weyl is an interesting example of the mood of uncertainty of the times. After joining the Brouwerian revolution in 1921, he moved back closer to Hilbert's views after the Hamburg address in 1927. And in the early 1930s he was for a short time Hilbert's successor in Göttingen.

International cooperation also suffered after the First World War. There were strong feelings on both sides. The *Conseil International de Recherches* was formed in 1919. Germany was excluded. Indeed, in the words of the French mathematician

Picard, Germany was a nation that had placed itself beyond humanity, "there is too much blood, and there are too many crimes, separating us". As an early example of the activities of the new *Conseil* Picard took an initiative to arrange an international congress of mathematics in Strasbourg in 1920. The choice of location was not arbitrary; the Germans were barred from attending. German scientists responded by calling their own *Naturforscherversammlung* in Nauheim. Brouwer attended and gave a lecture, *Does every real number have a decimal expansion?*, explaining the fine structure of the real numbers from an intuitionistic point of view.

The absurdity of boycotts and counter-boycotts could not continue. The situation inside Germany was not easy, should one accept the French-initiated *Conseil* as a satisfactory framework for international cooperation, or was a new start needed? Hilbert became a strong spokesman for re-establishing "normal" relationships within mathematics accepting the existing mechanisms. Brouwer argued that a new start was necessary and was in this respect more German (*Deutschfreundlich*) than most Germans themselves. When an international mathematics congress was organized in Bologna in 1928 with the cooperation of the *Conseil*, and the Germans were invited, he ended up by demanding a German boycott, much to the annoyance of Hilbert, who was invited to give the opening address at the meeting. There is an end to this story. The *Conseil* was dissolved and a new International Council of Scientific Unions was founded in 1931. It took longer time to bring order in the ranks of mathematicians, the present day International Mathematical Union dates from 1950.

The conflict between Hilbert and Brouwer had a sad end. Brouwer always had a high respect for the mathematician Hilbert, and Hilbert had around 1910 recognized Brouwer as the new star of topology and had, as noted above, tried to get him as the successor of the great Felix Klein in Göttingen. But *Grundlagenstreit* and *Deutschfreundlichkeit* led to a bitter end to their former friendship. The old master acted in an imperial fashion, Brouwer was dismissed from favor, in fact, he was dismissed as member of the editorial board of the Mathematische Annalen. This was at the time no small matter, and many, including Albert Einstein, were asked to intervene. But the old man could not be moved. But there were limits to his "victory". With Hitler came the collapse of Göttingen and German mathematics, and with Gödel came the collapse of Hilbert's foundational program as it was originally conceived. Gödel proved in 1931 his famous incompleteness theorem, which in particular implies that the consistency of arithmetics cannot be proved by finitary means. But whereas Hitler and his Germany forever have disappeared, proof theory in the Hilbert tradition prospered and is today a substantial part of constructive and computational mathematics.

In the1920s Brouwer published a number of papers on constructivism, in particular he obtained far-reaching results on the continuum and total functions. This contradicted classical mathematics and showed that Brouwer's program was incompatible with the traditional Hilbert approach. Constructivity in mathematics is, however, far more than mathematics in the style of Brouwer; we may point to the work by E. Bishop in his *Foundations of Constructive Analysis* and to P. Martin-Löf and his *Notes on Constructive Mathematics*. These studies develop a praxis of constructive mathematics matching in scope a large part of the "classical" theories. Of equal interest is the study of the philosophy and logic of constructive reasoning; a wide-ranging (but somewhat technical) study is the survey *Foundations of Constructive Mathematics* by M. Beeson. But whereas the praxis has had limited impact on how mathematics is done, the philosophy has flourished.

Brouwer was no logician, hence not interested in developing a formal system of axioms and proofs for his intuitionistic approach. For him intuition and structure was primary and, he would insist, independent of language and formalism. The early 1930s saw, however, the development of a formal system of intuitionistic logic. The pioneers were A. Kolmogorov in Moscow and A. Heyting, a student of Brouwer in Amsterdam. At the same time algorithms and computability were given a precise and general theory through the work of A. Turing. The community soon noticed the close connections between general algorithms and proofs in logical systems, in particular, in constructive proof theory, where existence means that you can actually "compute" an example. Again Martin-Löf is our example; extending the original Heyting logic he developed a powerful intuitionistic type theory as a framework for constructive mathematics. And he followed up with an influential lecture on *Constructive mathematics and computer programming* at the 1979 International Congress of Logic, Methodology and Philosophy of Science. His ideas are very much alive today. In the academic year 2012–2013 the Institute of Advanced Study in Princeton acted as host to a program in *Homotopy type theory*. This is, at first, a rather surprising link between homotopy theory, a branch of topology, and type theory and theoretical computer science. We shall not explain why, just quote the promise made by the organizers of the program of what is to be expected: "(The program) suggests a new conception of the foundation of mathematics, with intrinsic homotopical content, an "invariant" conception of the objects of mathematics—and convenient machine implementations, which can serve as a practical aid to the working mathematician".

We cannot know, but it would have been interesting to see how Brouwer, equally an expert in topology and constructivity, would have reacted to the fusion of types and homotopy. What we do know, is that deep insights will never disappear; Brouwer's achievements will be an integral part of a cultural heritage that will continue to enrich—in new and unexpected ways—our understanding of what knowledge is.

It is time to return to the "subtitle" and to discuss how mathematics is rooted in life. The subtitle is actually a quote from a letter Brouwer wrote to his thesis advisor in 1906 and is an early expression of how Brouwer saw mathematics as a free activity of the human mind and that mathematical truth was neither to be secured by reference to some abstract platonic realm, nor by reference to language, axioms and proofs of consistency. For Brouwer the creative activity of the individual mind was the focal point. There are also strong connections between Brouwer's philosophy and the phenomenology of Husserl. They met in Amsterdam in 1928 and had, as Husserl later wrote to Heidegger, long conversations which, in Husserl's words, "made a most significant impression on me". Brouwer on his side reported in a letter, "here, at the moment, Husserl is going around, I am also strongly attracted".

But with this exclusive focus on the individual mind there is a difficulty, shared with the formalists, of explaining the applicability of mathematics to the world around us. Mathematicians, as hired hands, are happy to apply their tools to analyze and sometimes solve an expanding set of problems about nature, society and mind. Mathematicians, as philosophers, have their difficulties, and there are a number of different schools of thought about the foundation of mathematics. It seems fair to say that neither Hilbert, nor Brouwer had a satisfactory and comprehensive philosophy of truth and applicability. We may take Hermann Weyl as our witness. In questions of philosophy and mathematical truth he sided with Brouwer. In his work on general relativity and quantum theory, he came quite close to Hilbert and his views on classical mathematics. But Weyl's long efforts to work out a unified philosophy of mathematics and its applicability remained unfinished. The *Grundlagenstreit* never produced a winner.

Perhaps, there is a way out. Between the individual mind and the platonic realm of structures outside time and space, there is culture, or—in the language of some anthropologists—there is the collective mind of the species. In a recent lecture, here reprinted as Chap. 9, *On what there is—infinitesimals and the nature of numbers* I have tried to argue for a cultural approach to the foundation of mathematics. Let me briefly quote one paragraph:

> P. Suppes has in a recent paper, *Why the effectiveness of mathematics in the natural sciences is not surprising*, discussed the emergence of geometry. He sees an early beginning in the perceptual processes necessary for humans to survive. He further notes in his paper how human culture developed the skill to represent "the structure of the external world in remarkable paintings and drawings", and concludes that the key to the early development of geometry lies in the gradually emerging "structural isomorphism between, in one case a perception and an object or a process in the world, and in another, between a mental image and an "abstract structure"". Modern geometry is thus a complex product of nature, mind and culture, and as such a "product" obviously applicable to what is seen ... .

At this point we should return to Brouwer. It is possible to see a connection between the quote above and Brouwer's views on how mathematics is linked to the outer world, see in particular his *Mathematik, Wissenschaft und Sprache* from 1929. The philosophers will, however, have to argue whether Brouwer in his 1929 paper is closer to Husserl's phenomenology than to a possible cultural foundation for mathematics.

In the introductory paragraph it was claimed that van Dalen's biography had an interest beyond the mathematical community. And I promised to explain why. The reader will judge if I have succeeded to present Brouwer, not only as a mathematician and philosopher, but also as a deeply engaged participant in the intellectual and political activities at a European level in the first part of the 20th century. I have also tried to trace the importance of his work for current research. The van Dalen biography runs to a total of 875 pages. In addition to the public life, it also tells the story of the private life of a strong, old-fashioned, and at times difficult professor. And for those who think of Brouwer in terms of dry and sometimes forbidding mathematical papers, his somewhat bohemian lifestyle in the little village Blaricum on the outskirts of Amsterdam may come as a surprise. The book is an engaging account of a great life.

References

The list of references contains a few items of possible interest to anyone who wants a deeper understanding of constructivity and its applications. The main reference is the book by van Dalen (2013), which is a rich and well documented source for the life and work of Brouwer and his time. This book has also a complete listing of all the published works of Brouwer. The classical textbook by Kleene (1952) is a thorough introduction to the ideas and the technical work of Hilbert, Brouwer, Gödel and Turing on foundational issues in logic and computing; some of their fundamental papers are reprinted in van Heijenoort (1967). Beeson (1985) and Troelstra and van Dalen (1988) are more recent and commendable updates. Feferman (1998) is recommended for its analysis both of Weyl's original work and of current ideas on predicativity in mathematics. Martin-Löf (1982) was an important step in linking constructive type theory and programming languages This has been carried forward in the current IAS project on homotopy type theory. The link between Brouwer and Husserl is discussed in van Atten (2007). Fenstad (2015), see Chap. 9 below, discusses some further philosophical issues related to the cultural approach to the foundation of mathematics.

Beeson, M. (1985). *Foundations of constructive mathematics*. Berlin: Springer.
Feferman, S. (1998). *In the light of logic*. New York: Oxford University Press.
Fenstad, J. E. (2015). On what there is—infinitesimals and the nature of numbers. *Inquiry, 15*, 57–79.
Institute for Advanced Study 2012–2013. *Homotopy Type Theory*. http://homotopytypetheory.org/book.
Kleene, S. C. (1952). *Introduction to metamathematics*. Amsterdam: North-Holland.
Martin-Löf, P. (1982). Constructive mathematics and computer program-ming. In L. J. Cohen et al. (Eds.), *Logic methodology and the philosophy of science VI*. Amsterdam: North Holland.
Troelstra, A. S., & van Dalen, D. (1988). *Constructivism in mathematics I and II*. Amsterdam: North-Holland.
van Atten, M. (2007). *Brouwer meets Husserl. On the phenomenology of choice sequences*, synthese library 335. Heidelberg: Springer.
van Dalen, D. (2013). *L. E. J. Brouwer: Topologist, intuitionist, philosopher*. Heidelberg: Springer.
van Heijenoort, J. (Ed.). (1967). *From frege to gödel: A source book in mathematical logic*. Cambridge, MA: Harvard University Press.

Part II
Language, Mind, and Numbers

Chapter 7
Tarski, Truth, and Natural Languages

Abstract The first part of this chapter traces the history of the relationship between logic and linguistics with particular emphasis on the contributions of the Polish logicians and philosophers of science A. Tarski and K. Ajdukiewicz. In the second part we give a brief review of current work on formal semantics for natural language systems and argue for the need for a richer geometric structure on the semantic model space.

Keywords Type theory—categorial grammar · Generalized quantifiers
Attribute-value formalism · Computational linguistics

The relationship between logic and language has always been a complex affair. They have been inseparably intertwined in the European intellectual tradition. It was not an easy relationship, but there were always issues of substance whether they were friends or foes. A somewhat personal view of this relationship was given in Chap. 5; here we shall limit our attention to the roles played by the Polish logicians A. Tarski and K. Ajdukiewicz in this development. We shall take the mid 1930s as our starting point and look at the situation from the perspectives of the Vienna Circle and the *Unity of Science* movement.

The Vienna Circle was a driving force behind the Unity of Science movement. A congress in Prague in 1934 had laid the plans for a series of future conferences devoted to the unity of science. An impressive International Committee had been formed to further these aims; we recognize the names of three Nobel prize winners in physics (Bohr, Bridgeman, and Perrin), several of the leading mathematicians of the time (e.g. Cartan, Fréchet, and Hadamard), and, of course, all the great names of the "new" philosophy, Carnap, Dewey, Lukasiewicz, von Mieses, Reichenbach, and Russell. There were some names from the social sciences, but almost none from the biological sciences. Physics, mathematics, logic, and "exact" philosophy were the standard bearers. This lead to the third congress in Paris 1937, which was devoted to the project of organizing an *International Encyclopedia of Unified Science*, where the foundation for science and rationality was once and for all to be inscribed; see Neurath (1955), one of the leaders of the movement, for a personal and very interesting account.

J. E. Fenstad, *Structures and Algorithms*, Logic, Argumentation & Reasoning 15,
https://doi.org/10.1007/978-3-319-72974-9_7

Linguistics was part of the unity of science project, and the well-known American linguist Leonard Bloomfield was to be its representative. The development of linguistics is an important, but sometimes a somewhat neglected part of our intellectual history. The study of languages covers a wide spectrum of topics, ranging all the way from the neural basis, via speech analysis and grammar formalisms, to the interpretation of metaphors. We shall be interested in the relationship between logic and linguistics.

Let us start by making a few remarks on the situation in linguistics in the mid 1930s. The historical and comparative school, which dominated by the end of the 19th century, had by that time been replaced by a structural approach going back to the great Swiss linguist Ferdinand Saussure. In the mid 1930s Bloomfield occupied a central position in the field. Philosophically he was close to the Unity of Science movement, being at the same time both a firm believer in the empiricism of the Vienna Circle and a committed behaviorist. This is very much in evidence in an article, *Linguistic Aspects of Science*, which he contributed to the new *International Encyclopedia of Unified Science*; see Bloomfield (1955). He is severely critical of the latin tradition in the study of grammar for the many mistakes caused by the "metaphysical", and hence un-scientific, definitions and a priori constructions found in this tradition. He recommended that linguistics as a science should seek its roots in the newly rediscovered grammar for Sanskrit. In this tradition we see, according to Bloomfield, a descriptive, even scientific, grammar rid of the metaphysics and the a priori approach of the Latin tradition. This set an example for the description of native Indian languages in the US and gave a historical justification for the data-driven and inductive methods of structural linguistics. If we believed in Bloomfield's account in the *Encyclopedia*, this would have been the final theory. But read on the background of current insight in the cognitive sciences, his behavioristic foundation for linguistics was a remarkably primitive construction.

Natural languages had, however, a very low status within the Vienna Circle. No one denied that it was something existing in space and time which had to be studied and classified by the proper (i.e. Bloomfieldian) methods. But it was not something to be used for insight and understanding. In the same *Encyclopedia*, devoted to the unity of science, we find the following programmatic declaration by C. W. Morris that "it has become clear to many persons today that man, including scientific man, must free himself from the web of words which he has spun and that language—including scientific language—is greatly in need of purification, simplification, and systematization"; Morris (1955). Tarski is an explicit source for Morris. And the specific source is, of course, the famous article *Der Wahrheitsbegriff in den formalisierten Sprachen*, reprinted in the collection Tarski (1956), where we read on page 165: "… the very possibility of a consistent use of the expression 'true sentence' which is in harmony with the laws of logic and the spirit of everyday language seems to be very questionable, and consequently the same doubt attaches to the possibility of constructing a correct definition of this expression".

The way out for the unity of science movement was to turn away from everyday language as a medium for scientific understanding and to construct a new and purified logical syntax and semantics for the "scientific man" of the future. It was left to R.

Carnap to be the architect for the new edifice, where his *Logische Syntax der Sprache* (Carnap 1937) was to be the central element of the new structure.

I do not know whether Bloomfield, Carnap and Tarski ever sat down together and discussed logic and language in the light of the doctrines of the *Einheitswissenschaft*. What I do know is that the future always likes to confound those who believe that the last battle has been fought and won. The interaction between logic and language did not at all develop as was prescribed in the *International Encyclopedia of Unified Science*. We recall a few facts.

American structural linguistics looked upon itself as the culmination of linguistics as a science. And the book that was intended as the last stone to the structure, was *Methods in Structural Linguistics* by Zellig Harris from 1951. The book was severely inductive in approach. Any description of a language must be based on an extensive occurrence analysis. The structure must emerge from observed regularities. General linguistics seemed to be reduced to a methodology for language description and experimental design. There is no universal grammar. And meaning is a stimulus-response afterthought. (I deliberately used the word "seemed" above; as explained in Chap. 5, there is much more to Harris than this crude descriptivism.) Such was the leading paradigm up to the appearance of Chomsky's *Syntactic Structures* in 1957. Chomsky was a revolution within linguistics, and no one can deny the importance of what he did. Much has been written about this. I shall add some remarks about the logic-linguistic interface.

Tarski denied that his logical analysis in *Wahrheitsbegriff* had any relevance for the understanding of truth and meaning in natural languages. This may be true. Tarski's analysis in *Wahrheitsbegriff* has remained a cornerstone in formal semantics; for a recent update on truth and partiality, see the author's article *Partiality* in the *Handbook of Logic and Linguistics* (1997). But if we are concerned with the broader interaction between logic and linguistics, we see a missed opportunity. What had happened if Tarski had paid attention to what his Warsaw colleague K. Ajdukiewicz tried to do in his article *Die Syntaktische Konnexität* from 1935? Tarski and Ajdukiewicz belonged to the same group of logicians in Poland in the mid 1930s. In the two papers mentioned above there are mutual references. Tarski has several references to Ajdukiewicz, also inside the section where he discusses "colloquial" languages (see footnote 1 on page 161 in Tarski 1956), but not to *Die Syntaktische Konnexität*. Ajdukiewicz has a reference to the *Wahrheitsbegriff* paper (see footnote 1 on page 209 of Ajdukiewicz 1967) where he notes the similarity between his theory of semantic categories and the hierarchy of logical types. Indeed, there are similarities. Both systems derive from the Russell-Whitehead theory of types; directly in Tarski's *Wahrheitsbegriff* and indirectly via Lesniewski (1929) in Ajdukiewicz's *Syntaktische Konnexität*. The link is—in hindsight—all too obvious. The categorial grammar of Ajdukiewicz is an applicative system corresponding to higher order logic. It was precisely for these logic systems that Tarski developed his concept of truth. Contrary to what both Tarski and Bloomfield (and their fellow travelers Morris and Carnap) maintained, it would have been possible in the 1930s to build a viable link between logic and linguistics. We could have had a modern version of the great synthesis of the medieval *modistae*; see the *Grammatica Speculativa* by Thomas of Erfurt from 1315. But both the methodological dogmas of the Vienna Circle and the strict inductive methodology of American structural linguistics forbade such speculations.

Not everyone followed the party line. Hans Reichenbach made use of higher order logic in a chapter on conversational languages in his text book *Symbolic Logic* from 1947, and in particular his analysis of the concept of time in natural languages has become a standard part of linguistic semantics. But there were no traces of a syntactic analysis in Reichenbach (1947). This we find in a paper by H. B. Curry, *Some Logical Aspects of Grammatical Structure*, from the late 1940s. Curry's paper was published in 1961 and was almost universally neglected by linguists. The breakthrough came first in the late 1960s with the work of Richard Montague; see the reprint of Montague's papers in Thomason (1974). Montague got his Ph.D. from Tarski in Berkeley, but that was on the metamathematics of set theory. I have no information if logic and natural languages ever was a topic of serious intellectual interchange between Tarski and his student.

Both Chomsky and Reichenbach-Curry-Montague worked at the interface between logic and linguistics. Chomsky entered the stage before semantics really became a standard part of the logic tool-box. He made use of the links between logic, formal languages, and computer science; in his work it is syntax which occupies the front stage. And since he does not have a workable theory of semantics, we can see how he "overloads" the syntactic component in his model building. Formal semantics came relatively late, a major influence was the Tarski school in Berkeley in the 1950s. Montague was a product of this school, so it was natural for him to have both syntax and semantics as formal parts of his model of (fragments of) natural language. Reichenbach as a very accomplished logician handled his semantics in a somewhat informal way, as was usual of the logicians of the early part of the 20th century; we may, as an example, recall how Skolem used an informal understanding of the semantics—even in his constructions of countable models of set theory and nonstandard models of arithmetic. Curry was more precise. He was a remarkably original logician, at times several steps in advance of his more well-known colleagues. Both in computer science and in theoretical linguistics he merits much greater recognition for his pioneering work.

Why was Montague and not Curry recognized for their work in linguistics? We can speculate. Curry worked in isolation and at odds with the ruling points of view. The time span from Harris in 1951 and Chomsky in 1957 is admittedly rather short, but it was soon recognized that structuralism had its limits—a theory of experimental design is not a theory of the subject matter itself—and that the new approach of Chomsky with its clean mathematical foundation and links to computer science was a promise for the future. Reichenbach and Curry with their insistence on logic and linguistics seemed to be a throw-back to an old-fashioned a priori approach and an unscientific metaphysics.

But Montague succeeded. The mathematics of Curry and Montague are comparable, and some may even prefer to read Curry rather than struggle with the formalism of Montague. But Montague had something which Curry did not—an interpreter. Barbara Hall Partee was—and still is—a leading American linguist. She was a professor in the Linguistic Department of the University of California at Los Angeles at the same time Montague was a professor of philosophy there, and she became acquainted with his work. She soon saw the relevance of it for the study of natural languages. She made his work accessible to the linguistic community and added significant new contributions herself. With her stamp of approval linguists had to take note.

7.1 Logic and Grammar

It is time to fill in some technical details. The theory of grammar was never a unified science. There seems, however, to be one shared assumption. Most theoretical linguists will subscribe to the following point of view expressed by the Dutch linguist Jan Koster in a lecture some years ago; Koster (1989):

> … we have to make a distinction between a computational structure and a more or less independent conceptual structure … The conceptual module includes what is often referred to as knowledge of the world, common sense and beyond. It also includes logical knowledge and knowledge of predicate-argument structure. The computational module concerns the constraints on our actual organization of discrete units, like morphemes and words, into phrases and constraints on relation between phrases. (Koster 1989, p. 593)

Later he adds (Koster 1989, p. 598)—and this is not a view universally accepted by linguists:

> If the two basic modules are autonomous and radically different in architecture and evolutionary origin, there must be … an interface … My hypothesis is that the interface we are looking for is the lexicon ….

A major part of post-Chomsky linguistic theory has been devoted to the investigation of the computational module. Post Montague we have seen an increased interest in the conceptual module, or semantic space. In a certain sense the computational module of the linguist corresponds to syntax as used by the logician and the conceptual module corresponds to the use of semantics or model concept. This is a rough indication and the usage may differ between different authors.

Both in Curry (1961) and Montague (1974) we have two major parts to the model. One is an *extended categorial grammar*, corresponding to the computational module of Koster, which gives a syntactic analysis of a well-defined fragment of a natural language. The second part, which corresponds to the conceptual module of Koster, consists of a *translation* into a system of higher order (intensional) logic. The translation is required to observe the principle of *compositionality*, which means that it is (in an appropriate sense) a *homomorphism* between the syntactic "algebra" and the semantic "algebra".

In his analysis Montague implicitly used the notion of *generalized quantifier* to give a uniform treatment of *noun phrases*; see Fenstad (1978) and Barwise and Cooper (1981). Thus if *john* and *mary* are items of the lexicon, they will not be translated as constants j and m of individual type, but will be translated as:

john translates to $\lambda P.P(j)$

mary translates to $\lambda Q.Q(m)$

The term $\lambda P.P(j)$ corresponds in the semantic interpretation to the set $\{X: j' \varepsilon X\}$ in the model, where j' is the value of the constant j. Note that for simplicity we use an extensional version of higher order logic.

The sentence *john loves mary* has in the Montague model the following syntactic analysis:

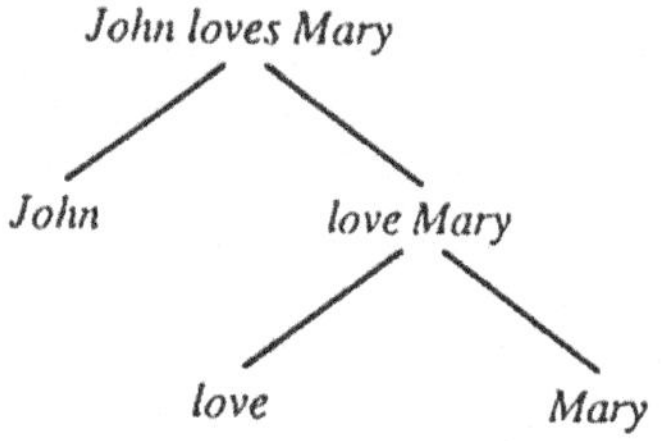

Compositionality, or the homomorphism criterion, demands a translation of *love mary* as an application of the translation of *love* to the translation of *mary*,

love mary translates to $love'(\lambda Q.Q(m))$,

where *love′* is an appropriate constant of the logic as translation of the lexical item *love* of the natural language fragment. In the next step compositionality dictates the following translation:

john loves mary translates to $(\lambda P.P(j))(love'(\lambda Q.Q(m)))$.

Compositionality requires that every node in the syntactic tree have an associated object in the logic and the semantic algebra. And the syntactic analysis tells us which type to assign to the associated object. Thus the primary translation *love′* is no longer a relation between individuals, but a more abstract function. Not everyone in love would be happy with this analysis!

There is an answer that some people have taken to be a success of the Montague analysis; see Fenstad (1978). The logic may have a basic constant *love** denoting a relation between individuals. Using lambda-abstraction we can then define a new constant *love′* of the appropriate type by

$$love'(\mathbf{P})(x) \quad \text{if and only if} \quad \mathbf{P}(\lambda y.love^*(x,y))$$

This "lifted" *love′* has the correct type to match the syntactic analysis:

$$\begin{array}{lll} love'(\lambda Q.Q(m))(j) & \text{conv. to} & (\lambda Q.Q(m))(\lambda y.love^*(j,y)) \\ & \text{conv. to} & (\lambda y.love^*(j,y))(m) \\ & \text{conv. to} & love^*(j,m) \end{array}$$

This is what Tarski and Ajdukiewicz could have done in the mid 1930s. Everything was in Curry (1961), a paper which was written in the late 1940s, but largely neglected. We had to wait for Montague in the late 1960s to see the modern equivalent of Thomas of Erfurt and the revival of the link between logic and linguistics. It is outside of the scope of this account to follow the more recent developments. A good source for this is the *Handbook of Logic and Language* (van Benthem and ter Meulen 1997). The reader curious for more details on type theory and its applications to linguistics may also consult Carpenter (1997).

The notion of *the connecting sign* was a central element of Koster's analysis. And, indeed, the nature of the *syntax-semantics interface* has been a major issue in recent linguistic theory. In early Chomsky (see his *Aspects of the Theory of Syntax* from 1965) the connecting sign between grammatical deep structure and semantical representation was basically an arrow decorated with a name (a so-called projection rule)—but with no particular content; we noted above that formal semantics was not well developed at the time when Chomsky started his work. With Montague, a student of Tarski, semantics is part of the tool-box, and, as we have seen above, in his model the connecting sign is most readily identified as the translating formula in higher order intensional logic.

A major alternative to formulas as connecting signs is the use of attribute-value structures. This is an approach which in a very concrete sense uses a lexical sign as the connecting element between the computational and conceptual modules. An early and important example of such a theory is the *Lexical-functional grammar* (LFG) developed by Kaplan and Bresnan (1982). This format was extended to include semantical attributes in *Situations, Language and Logic* (Fenstad et al. 1986). A closely related theory is the system of *Head-Driven Phrase Structure Grammar* (HPSG) developed by Pollard and Sag (1987). *The Discourse Representation Structures* theory developed by H. Kamp (see the exposition in Kamp and Reyle 1993) gives another example of a theory of the connecting sign bearing some similarity to attribute-value theories. And within AI we have seen a number of proposals in the form of trees, frames, scripts, etc.

Every theory of language has as its ultimate goal an account of the link between linguistic structure and meaning. But different theories differ as to where to locate the connecting sign along the syntax-semantics axis. Chomsky's *Aspects* is biased toward the syntax end. To a certain extent this is also true of classical LFG. HPGS is more balanced between the two components. From our point of view Montague's theory is above all a theory of the connecting sign. The grammar is simple and the theory makes almost no use of the model structure. Much early theory in the Montague tradition consisted in the manipulation of lambda-terms. At this point we should remind the reader that the *Situation Semantics* developed by Barwise and Perry (1983) has emerged as a major alternative to the standard Tarski-type semantics;

see several articles in the *Handbook of Logic and Language* (van Benthem and ter Meulen 1997). It was also a system of situation semantics that was used in Fenstad et al. (1986) to add a meaning component to LFG; see the paragraph above.

Cognitive Grammar is an interesting exception to this tradition. Here we see a shift from the syntax to the semantics end of the connecting axis; for a development of this theory see Langacker (1987, 1991). In the earlier forms of the theory we still recognize two components, one phonological and one semantical. There is also a connecting symbolic structure, which in some sense is a pictorial attribute-value structure. In a purer version of the theory the phonological module is subsumed under the semantic one: "... phonological space should instead be regarded as a subregion of semantic space (Langacker 1987, p. 78)". Cognitive grammar is a minority view in linguistics, but it may have a valid point in its revolt against the combinatorial dominance within current linguistic theory; for a discussion of this point and the geometrization of thought see Gärdenfors (2000).

Contrary to cognitive grammar the standard view in linguistics has been that syntax is the only input to semantic interpretation. This is particularly true of categorial grammar and theories in the Montague tradition which are based on type-theoretic formalisms. In these theories meaning is usually seen as a homomorphic image of the syntax, and this in a very precise sense: the interpretation function is a homomorphism (i.e. a structure preserving) map between the algebra of formulas and the set-valued algebra generated by the model as explained in the example above. The world is what can be expressed in your language.

We can read a different philosophy behind the attribute-value formalism. This formalism can be looked upon as a constraint-based view of the relationship between linguistic form and meaning.

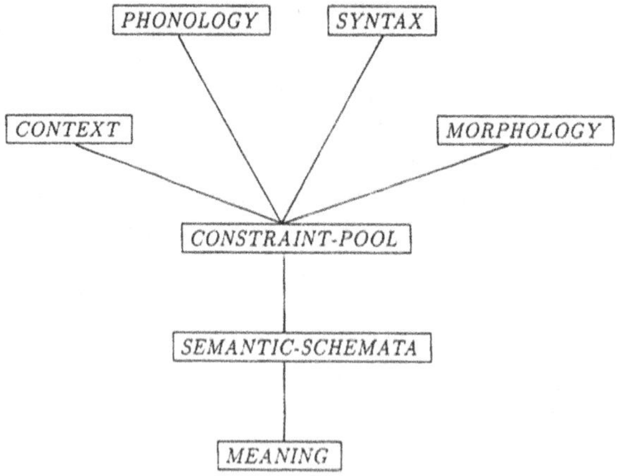

In this view all aspects of linguistic form, such as phonology, syntax, and morphology as well as context contribute to a combined set of constraints, which in turn determines the meaning of an utterance. In theories of the LFG variety the constraints are given in form of equations and the resulting linguistic sign or attribute-value matrix represents a consistent solution to the constraint equations; for an extended discussion see Chap. II of *Situations, Language and Logic* (Fenstad et al. 1986).

Let us pause for a moment and reproduce the canonical LFG example of a linguistic sign. The sentence is

a girl handed the baby a toy

Following tradition we simplify and look at syntax alone as the source for constraint equations. In this case the attribute-value matrix is:

$$\begin{bmatrix} \text{SUBJ} & \begin{bmatrix}\text{SPEC} & \text{A}\\ \text{NUM} & \text{SG}\\ \text{PRED} & \text{'GIRL'}\end{bmatrix} \\ \text{TENSE} & \text{PAST} \\ \text{PRED} & \text{'HAND } \langle(\uparrow\text{SUBJ})\ (\uparrow\text{OBJ2})\ (\uparrow\text{OBJ})\rangle\text{'} \\ \text{OBJ} & \begin{bmatrix}\text{SPEC} & \text{THE}\\ \text{NUM} & \text{SG}\\ \text{PRED} & \text{'BABY'}\end{bmatrix} \\ \text{OBJ2} & \begin{bmatrix}\text{SPEC} & \text{A}\\ \text{NUM} & \text{SG}\\ \text{PRED} & \text{'TOY'}\end{bmatrix} \end{bmatrix}$$

This is a very simple example and tells us only that the theory is pointed in the right direction. There are various noteworthy points. Let me just draw attention to the attributes SUBJ, OBJ, and OBJ2. In one direction they represent a noun phrase, in another a generalized quantifier—the link between the two is, in fact, a true link between language and logic. As pointed out above this link is implicit in Montague (1974), but it took some time to excavate this fact from his formalism.

The matrix above is a lexical sign. A sentence such as *a girl handed the baby a toy* gives us a complete sign, a verb phrase alone would determine an incomplete sign where only the attributes PRED and TENSE would have values, but the sign calls for values of the three other attributes SUBJ, OBJ and OBJ2. In simple situations like this one everything works out perfectly. But when the theory is called upon to serve in more complex situations, where context is important for understanding, we need a more elaborate structure theory for lexical signs. This is a focus for much work today and an extended theory for lexical signs is an absolute necessity for viable language engineering applications.

A useful review of current work is given in R. Blutner *Lexical semantics and pragmatics* (2002). After discussing the defects of current theories Blutner points to two ways out. One way is represented by a theory of *The Generative Lexicon* as developed by Pustejovsky (1995). This approach aims to enrich the lexicon with new generative mechanisms while staying as close as possible to current linguistic technology. The advantage of this is that it keeps the changes close to the existing computable structures and is therefore computationally tractable. Pustejovsky's presentation of his theory is not easy to understand and leaves some doubts to the technical coherence of parts of his work. In his Oslo thesis, *Leksikalsk semantikk,* Skrindo (2001) has presented part of the theory in a consistent way within the framework of *Situations, Language and Logic* (Fenstad et al. 1986). Skrindo's thesis also gives a useful review of much current work.

Blutner argues against the generative position adopted by Pustejovsky and presents a pragmatic alternative, pointing to the need for mechanisms of contextual enrichment (i.e. a pragmatic strengthening based on contextual and encyclopedic knowledge). But much remains to be done to translate this view into a workable theory and efficient computational praxis.

We remarked above that much of current theory is a theory of the connecting sign and that little attention is paid to the structure of semantic space. In Montague's theory the semantic model is there, but it is almost never used. In Pustejovsky's theory one is still close to the attribute-value matrix format. There is a model structure, but not more than can be written into the representational form. Seen from the point of view of language engineering this is not necessarily a criticism. Any application—take any question-answering system as an example—reduces necessarily the semantics to a data base representation.

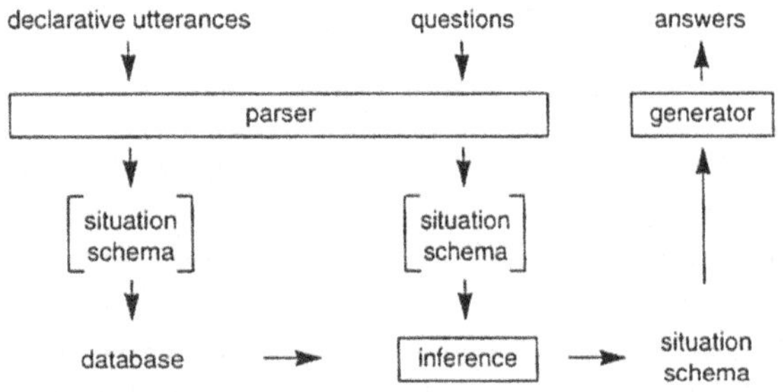

This example of an question-answering system is taken from Vestre (1987) and is quite typical in overall structure for applications where "model = data base" is a valid equation. This application is further discussed in Chap. 8.

Formal semantics, which is a name used for much of the work at the interface between logic and language, has in my view been a major influence on theoretical linguistics. But there are limits. Computational linguistics never was and is not an applied branch of theoretical linguistics. There are many useful links and the equation "model = data base" is a useful starting point, but computational linguistics and language engineering need to draw upon many sources. To be very brief and somewhat polemical, when you are given the task to design a dictation system for journal keeping to be used by doctors in the health service system, you do not start out by consulting a manual of higher order intensional logic. This is a situation not very different from the traditional relationship between basic physics and, say, electrical engineering. Few, if any, actual design tasks would start from the Maxwell equations. There have, however, been important changes in the physics-engineering relationship. In many parts of material technology, applications today are in a very direct sense applied science. Will we see a similar development in computational linguistics and language engineering? We recommend two collections of papers for further reflection, *Computational linguistics and formal semantics* (Rosner and Johnson 1992) and *Computers, language and speech: formal theories and statistical data* (Spärck Jones et al. 2000).

We have noted some limits in the relationship between the science of language and language technology. There are also major challenges in the relationship between the science of language and the broader cognitive sciences; for an overview see the next chapter, *Formal semantics, geometry and mind* and P. Gärdenfors' book, *Conceptual spaces: The geometry of thought* (Gärdenfors 2000). Formal semantics must proceed beyond the equation "model = data base" in order to serve as a link between language and mind. The challenge is to understand how language and cognition are rooted in the behavior of large and complex assemblies of nerve cells in the brain. Much of current linguistic theory proceeds at the level of symbol manipulation, such as sorting, ordering and comparing symbols. But, as was implicit in Blutner's discussion of pragmatics and the lexicon, we need to proceed beyond this level. A first step is a geometrization of model theory as a basis for a phenomenological model of mind. The next is to understand how the dynamics of brain cell interaction generates this geometry. Much is already known, and some of this will be told in the next chapter, but much remains before we have a firm understanding of this double task.

References

Ajdukiewicz, K. (1967). Syntactic connection. In S. MacCall (Ed.), *Polish logic* (pp. 1920–1939). Oxford: Oxford University Press. (This is a translation of the original article Die syntaktische Konnexität from 1935).

Barwise, J., & Cooper, R. (1981). Generalized quantifiers and natural languages. *Language and Philosophy, 4,* 159–219.

Barwise, J., & Perry, J. (1983). *Situations and attitudes*. Cambridge, MA: The MIT Press.

Bloomfield, L. (1955). Linguistic aspects of science. In O. Neurath et al. (pp. 219–277).

Blutner, R. (2002). Lexical semantics and pragmatics. *Linguistische Berichte, 10,* 27–58.

Bresnan, J. (1982). *The mental representation of grammatical relations*. Cambridge, MA: The MIT Press.

Carnap, R. (1937). *Logical syntax of language*. London: Routledge and Kegan Paul. (This is a translation of the German original *Logische Syntax der Sprache* from 1934).

Carpenter, B. (1997). *Type-logical semantics*. Cambridge, MA: The MIT Press.

Chomsky, N. (1957). *Syntactic structures*. TheHague: Mouton and Co.

Chomsky, N. (1965). *Aspects of the theory of syntax*. Cambridge, MA: The MIT Press.

Curry, H. B. (1961). Some logical aspects of grammatical structure. In R. Jakobson (Ed.), *The structure of language and its mathematical aspects*. Providence, RI, USA: American Mathematical Society.

Fenstad, J. E. (1978). Models for natural languages. In J. Hintikka, I. Niiniluoto, & E. Saarinen (Eds.), *Essays on mathematical and philosophical logic*. Dordrecht: D. Reidel Publishing Company.

Fenstad, J. E., Halvorsen, P. K., Langholm, T., & van Benthem, J. (1986). *Situations language and logic*. Dordrecht: D. Reidel Publishing Company.
Gärdenfors, P. (2000). *Conceptual spaces: The geometry of thought*. Cambridge, MA: The MIT Press.
Harris, Z. (1951). *Methods in structural linguistics*. Chicago: The University of Chicago Press.
Kamp, H., & Reyle, U. (1993). *From discourse to logic*. Dordrecht: Kluwer Publishing Company.
Kaplan, R. & Bresnan, J. (1982). *Lexical-functional grammar*. In J. Bresnan (1982), (pp 173–281).
Koster, J. (1989). How natural is natural language. In J. E. Fenstad, I. T. Frolov, & R. Hilpinen (Eds.), *Studies in logic* (Vol. 126, pp. 591–606). Amsterdam: North-Holland Publishing Company.
Langacker, R. (1987, 1991). *Foundation of cognitive grammar* (Vols. I and II). Stanford, CA: Stanford University Press.
Lesniewski, S. (1929). Grundzüge eines neuen Systems der Grundlagen der Mathematik. *Fundamenta Mathematicae, 14,* 1–81.
Montague, R. (1974). The proper treatment of quantification in ordinary english. In R. Thomason (pp. 247-270).
Morris, C. W. (1955). Foundation of the theory of signs. In O. Neurath et al. (pp. 77–137).
Neurath, O. (1955). Encyclopedia and unified science. In O. Neurath et al. (pp. 1–27).
Neurath, O., Carnap, R., & Morris, C. (Eds.). (1955). *International encyclopedia of unified science* (Vol. I). Chicago: The University of Chicago Press.
Pollard, C., & Sag, I. (1987). *Information-based syntax and semantics* (Vol. 1). Stanford, CA: CSLI.
Pustejovsky, J. (1995). *The generative lexicon*. Cambridge, MA: The MIT Press.
Reichenbach, H. (1947). *Elements of symbolic logic*. London: Macmillian.
Rosner, M., & Johnson, R. (Eds.). (1992). *Computational linguistics and formal semantics*. Cambridge: Cambridge University Press.
Skrindo, K. (2001). *Leksikalsk Semantikk (in Norwegian)*. Oslo, Norway: Matematisk Institutt.
Spärck Jones, K. I. B., Gazdar, G. J. M., & Needham, R. M. (Eds.). (2000). *Computers, language and speech: Formal theories and statistical data*. London, UK: The RoyalSociety.
Tarski, A. (1956). The concept of truth in formalized languages. In A. Tarski (Ed.), *Logic, semantics, metamathematics*. Oxford: Oxford University Press. (This paper is a translation of the original article *Der Wahrheitsbegriff in den formalisierten Sprachen* from 1936).
Thomason, R. (1974). *Formal philosophy selected papers of Richard Montague*. New Haven, CT: Yale University Press.
Thomas of Erfurt. (1972). *Grammatica speculativa* (G. L. Bursill-Hall, Ed. and Trans.). London: Longman.
van Benthem, J., & ter Meulen, A. (Eds.). (1997). *Handbook of logic and language*. Amsterdam: North Holland Publishing Company.
Vestre, E. (1987). *Representasjon av Direkte Spørsmål (in Norwegian)*. Oslo, Norway: Matematisk Institutt.

Chapter 8
Formal Semantics, Geometry, and Mind

Abstract Standard theory of grammar postulates the existence of two modules, one being a conceptual module which includes what is often referred to as knowledge of the world, one being a computational module which is concerned with the constraints on our actual organization of discrete units, such as morphemes and words into phrases. Much of current theory is a theory of the syntax/semantics interface, i.e. a theory of how to connect grammatical space (the computational module) with semantical space (the conceptual module). In addition there has always been much work on the structure on grammatical space. However, remarkably little work has been devoted to the structure of semantical space. Even the Montague grammarians rarely make any use of the structure of their models; working almost always at the level of lambda-terms. It is now time to add geometry to this picture.

Keywords Formal semantics—ontology · Conceptual spaces
Geometry and mind · Common nouns—attractor networks

A major part of current linguistic theory has been focused on the investigation of grammatical space. There is no need to review this work here; I assume that the reader is familiar with current representational forms, either as tree-like structures or as attribute-values structures. There is a continuing discussion of how many subunits one needs to recognize within the combinatorial component—of phonological, morphological and/or syntactic nature—and how to determine their exact interdependence; this is not our concern in the present discussion. As to the nature of the conceptual component I shall at this point only make one preliminary assumption, that it has—at some level—a standard model-theoretic core. This claim is not uncontroversial; it is one of the aims of this chapter to argue that model theory—correctly understood—is a necessary link between grammar and mind.

It remains to add a few comments on current theories of the syntax-semantics interface. There exists a large and sometimes bewildering number of proposals. In the framework proposed by Chomsky in *Aspects of the Theory of Syntax* (see Chomsky 1965) the connecting sign between grammatical deep structure and semantical representation was basically an arrow decorated with a name (a so-called projection rule)—but with no particular content. In R. Montague's theory as presented

J. E. Fenstad, *Structures and Algorithms*, Logic, Argumentation & Reasoning 15,
https://doi.org/10.1007/978-3-319-72974-9_8

e.g. in his *Proper Treatment of Quantifiers in English* (see Montague 1974), the connecting sign was a logical formula in a system of higher order intentional logic. The connecting sign may also take the shape of an attribute-value structure. *Lexical-functional grammar* (LFG) is an important example of such a theory (see Kaplan and Bresnan 1982). This format was extended to also include semantical attributes in the work on situation semantics; see the exposition in *Situation, Language and Logic*, Fenstad et al. (1987).

```
[ REL      marry                                ]
[ ARG. 1   [ IND    John ]                      ]
[          [ IND    IND. 1                    ] ]
[          [ SEPC   a                         ] ]
[ ARG. 2   [        [ REL      girl   ]       ] ]
[          [ COND   [ ARG. 1   IND. 1 ]       ] ]
[          [        [ POL      1      ]       ] ]
[          [ IND    IND. 2                    ] ]
[ LOC      [        [ REL      <      ]       ] ]
[          [ COND   [ ARG. 1   IND. 2 ]       ] ]
[          [        [ ARG. 2   IND. 0 ]       ] ]
[ POL      1                                    ]
```

The situation schema is a representational form computable from linguistic structure; in this case from the sentence *John married a girl*. One key point of the analysis is to rewrite a noun phrase as a generalized quantifier; see the ARG.2 attribute. We also see how the tense marker PAST of traditional LFG analysis is rewritten as a space-time attribute LOC; this is of crucial importance in the analysis of locative prepositional phrases (see Colban 1987).

A further example of the use of an attribute-value structure as a connecting sign is the theory of *Head-Driven Phrase Structure Grammar* developed by C. Pollard and I. Sag (see Pollard and Sag 1987). H. Kamp's theory of *Discourse Representation Structures* (see the recent exposition in Kamp and Reyle 1993) gives another example of a connecting sign bearing some similarities to attribute-value theories. Within AI research we have seen a variety of proposals in the form of trees, frames scripts, etc. We need not enter into further details concerning the syntax of the connecting sign since our standing assumption is that at some level there is a model-theoretic core; the structure of this core is our concern.

Every theory of language has as its ultimate goal an account of the link between linguistic structure and meaning. Some theories, e.g. Chomsky's model in *Aspects*, are biased toward the combinatorial component; others, such as the HPSG account, are more balanced between the combinatorial and conceptual modules. From our point of view Montague's theory is primarily a theory of "the connecting sign"; his grammar is simple and he makes remarkable little use of the structure of his semantical component; the discussion can in almost all cases be carried out at the level of lambda-terms. The model theory is in an indirect sense necessary; it allowed him to draw several relevant linguistic distinctions, e.g. between a *de re* and a *de dicto* reading of noun phrases.

Cognitive grammar is an interesting exception. Here we see—against Chomsky—that the center of theoretical gravity has moved towards the conceptual end of the theoretical spectrum. The theory has been extensively developed by R. W. Langacker (see his two volumes *Foundation of Cognitive Gramma*r 1987, 1991). In its original form the theory still recognizes two modules, one phonological and one semantical. In addition there is also a connecting symbolic structure, which is closely related in form to an attribute-value matrix. But in more extreme versions of the theory we see that the phonological component is subsumed under the semantic one; "... phonological space should instead be regarded as a subregion of semantic space (Langacker 1987, p. 78)".

We may have some sympathy with the cognitive grammarian's revolt against the combinatorial dominance within current theoretical linguistics. But at the same time we are surprised to see the rather primitive formal apparatus used by Langacker. There seem at first to be few connections between the hand-drawn representations in *Foundation of Cognitive Grammar* and other parts of cognitive science. We shall return to this issue in our discussion of conceptual spaces. Our first and preparatory task, however, will be a critical review of standard logical semantics and its ontology. This is necessary—even urgent—since we have already proposed that model theory is the "core" of any theory of natural language understanding.

8.1 Formal Semantics and Its Ontology

First order logic—seen from a proof-theoretic perspective—is a system of remarkable strength. It has a complete proof procedure, i.e. every universally valid formula is provable. And all of mathematics, dressed in its set-theoretic garb, is formalizable within the system. Yet this strength is illusory; we shall explain why.

Logic, as well as language in general, has two sides; one is syntax and proofs, the other is semantics and validity. In first order logic we seem to have a perfect balance between the two. On the syntactical side we have notions such as proofs and theorem, on the semantical side we have the notions of model and validity. These notions are in the general theory of first order logic perfectly matched through the celebrated Gödel completeness theorem: a formula is provable, i.e. is a theorem of first order logic, if and only if it is universally valid, i.e. true in all models. Gödel's theorem is primarily a technical result, but it also is an insight with broader explanatory power.

In a sufficiently wide sense, logic is a natural science: We have a pre-formal idea of truth and, hence, of a correct argument. Logic converts this pre-formal idea into two distinct technical notions, the syntactical notion of proof and the semantical notion of truth in a model. Informally we may convince ourselves that what is provable is necessarily true in the pre-formal sense, and what is true in the pre-formal sense is, in particular, true in all models. Thus the pre-formal sense of truth is caught between the two technical notions of proof and truth in a model. Gödel's completeness theorem, which asserts that truth in all models implies provability, closes the circle and seems to show that the pre-formal notion of truth has a correct and adequate analysis in terms of the technical notions of proof and truth in all models. This is an exemplary piece of applied science; a natural phenomenon has been given a sound and comprehensive theoretical analysis. We can go one step further. The notion of proof, being a finite combinatorial structure, is algorithmic. The notion of being provable is not: a formula is provable if and only if it has a proof—but we cannot effectively decide in general if a proof exists or not. However, fragments of logic have effective proof procedures; it is a valid scientific strategy to explore the limits of effective computability and to combine the search for algorithms with various heuristic and probabilistic recipes. From this perspective reasoning and understanding seems to be reduced to the search for ever more sophisticated proof procedures. But, despite the successes, the strength is illusory.

The reason is ontological. The true ontology of first order logic is an ontology of lists; we explain why. A model for first order logic is a set-theoretic structure consisting of a non-empty set, the domain A, and a collection of relations, $R_1, R_2, \ldots$, defined on the domain A. We use the word set-theoretic to emphasize that an n-ary relation at this level of explanation is nothing more than a set of n-tuples over the domain. An n-ary relation on A can be represented in the form of two lists. One a list of basic or atomic positive facts, i.e. a list of all

(i) $$Ra_1 \ldots a_n$$

such that $(a_1, \ldots, a_n)$ belongs to the relation R; and a supplementary list of basic or atomic negative facts,

(ii) $$not - Rb_1 \ldots b_n$$

for $(b_1, \ldots, b_n)$ not belonging to the relation R. The lists which may be infinite if the domain A is infinite, are complete, i.e. any n-tuple over A defines an item in either list (i) or (ii). The important thing to recognize is that the model structure is "flat". All objects of A have the same ontological status, there are no structurally defined hierarchies. Of course, a relation R may in a specific structure impose a hierarchy on its domain. But to prove deductive completeness theorems we must necessarily speak of all models. This means, e.g. in applications to set theory, that the semantical interpretation is a coding of a possibly rich structure into a flat domain. This may be adequate for book-keeping purposes, even for linguistic engineering applications, but not for understanding, in particular, for natural language understanding.

This limitation of first order logic has always been recognized, and several extensions have been studied. We mention briefly:

higher types and partiality,
possible worlds and situations,
order/time/events and episodic logic,
masses and plurals.

The literature on extensions of first order logic is vast; we can only mention a few references of special relevance for our current purpose. Partiality and higher types is reviewed in Fenstad (1996). Situation theory, partiality and the relational theory of meaning is the topic of Barwise and Perry (1983). The theory of mass nouns and plurals have been studied by G. Link and J. T. Lönning; for general references see Lönning (1989).

The extensions are, however, "tame"; all of them rest on the ontology of lists. A partial model corresponds to partial lists, which means that if the relation R in a structure A is partial, then there are n-tuples over A which occur neither in (i) nor in (ii) above. In a similar way structures of higher types correspond to lists of lists, and possible world structures correspond to collection of indexed lists. A small extension occurs with order and time. In these cases the domain A has a fixed order relation, but since theories of linear, partial, and other types of order are algebraic and easily axiomatized, we remain with the same ontology. We should be careful to note that this ontology is adequate and productive for many technological applications. We have recently seen a fruitful merging of ideas from data base theory and logic. This is not the place to review current advances in relational and deductive data base theory and to see how data base theory is becoming more and more intertwined with finite model theory. We shall, however, indicate the connection by presenting a simple example of a Q-A (i.e. question-answer) system.

The system has the following architecture (Vestre 1987; see also the exposition in Fenstad et al. 1992).

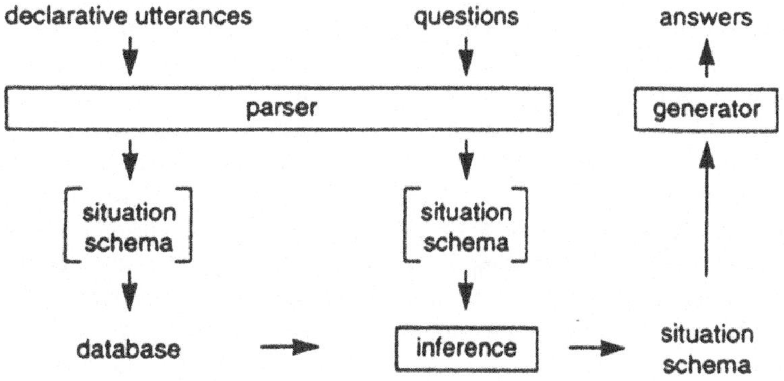

The system stores factual information in a data base. This data base can be updated by accepting input sentences within a natural language fragment. The parser rewrites such sentences in the format of a situation schema, which is a particular kind of an attribute-value structure (Fenstad et al. 1987). The system automatically extract from the attribute value structures basic facts—positive or negative—and stores these facts in the data base. A question is rewritten as an incomplete situation schema. The inference module accepts these incomplete structures and tries deductively to transform the incomplete sign into a complete attribute-value structure, using information from the data base as admissible hypotheses for the proof. Finally, the generator transform the completed sign into a grammatically correct answer. The architecture of the system is open. The system can be extended to accept a larger class of sentences and questions by extending the syntactical component. The domain of application can be extended by reading into the system more information. The system can also be made more efficient by further development of the inference subpart. Seen in this perspective the system is successful as an example of language engineering. But there are limitations.

A sufficiently precise question will receive a correct and exhaustive answer. But if the question is more open and searching, it may signal that we are interested in answers which are "relevant" and "informative" rather than just correct in a strict logical sense. This may be the case when a question is part of a larger dialogue situation. An adequate analysis of dialogues is, however, not a matter of grammar alone. If this point is granted, we immediately see the limitations of the system, viz. the conceptual module, i.e. meaning and understanding, is reduced to a data base which is nothing more than a list of basic facts. The equation "model = data base" has useful technological applications, but it cannot serve as a basis for an analysis of the flow of meaning in a natural language dialogue.

To analyze the situation in some more detail we need to review a few facts from cognitive development theory. This is a domain of many theories, often at odds with each other. What we need will only to a small degree be touched by these controversies. With some justification we shall therefore concentrate on only one account—the story as told in M. Donald, *Origin of the Modern Mind* (Donald 1990, 1993).

According to Donald we can recognize four stages in the evolution of culture and cognition: *episodic culture*, *mimetic culture*, *mythic culture* and *theoretical culture*. The episodic culture was characterized by the ability to react to a fixed situation in which the individual was placed. In a certain sense one "understood" the significance of this situation and could react accordingly. But this insight could not be used to understand when placed in other analogous situations. The transition to mimetic culture extended the limited mode of understanding found in episodic culture. The individual was now able to decompose the meaning of a specific situation into components and to recombine these in order to develop effective patterns of reaction in other, but similar situations. A rudimentary communication system between the members of a tribe also developed. According to Donald this was pre-linguistics and exclusively composed of gestures and the mimetic elements. The transition from mimetic to mythic culture was above all characterized by the acquisition of language.

We now see an emerging communication system—a possibility for dialogue—which is a combination of gestures and speech. Donald is very careful to ground his theories in anthropological data on anatomical and cultural development. He stresses, in particular, how the development is cumulative, e.g. in a dialogue we have both mimesis and speech, the latter stage has not suppressed the former—a dialogue is not reducible to a text.

In the third transition, from mythic to theoretical culture, there are no anatomical changes. We see a transition to a stage where we have access to external symbolic storage. The development of written language was an early stage in this transition; present-day information technology represents another. For the philosophically informed it may be instructive to compare Donald's analysis of the third transition to Popper's theory of the so-called World III (see Popper 1972); another and compatible story is the anthropological analysis of culture (see White 1947). There are many comments one could want to make; I shall only emphasize one crucial new element at stage four, viz. the possibility of an external memory without an anatomical foundation, or—expressed in a more colorful language—the possibility of a "collective mind" independent of the individual brains of the species. This explains why certain applications of linguistics to engineering are successful and why others fail.

A limited Q-A system can be completely located within the fourth cultural stage, i.e. theoretical culture with the possibility of external symbol storage. None of the previous stages need to be involved; semantical space is therefore reducible to lists and data bases, as was the case in the system described above. But a real-life dialogue lives at the intersection of many cultures, in particular, the mimetic and the mythic. The dialogue is a combination of gesture and speech and is therefore not reducible to an external text. The dialogue is seen to pose other and more intractable challenges to the conceptual module; the successful analysis of dialogues, beyond a few stereotypes, is not a matter of incremental extensions of current linguistic technology. On the other hand, limited translation systems typically belong to the fourth stage, which means that an ontology for translation is an ontology of external objects, and, hence, reducible to an ontology of lists and lexicons; for an interesting example of a limited and interactive translation system see the PONS system developed by Dyvik (1993). The situation will be radically changed if we want to build a system for combined speech and vision. As a starting point we may be looking for something very simple—e.g. a system which shall recognize objects simply by name. To simplify even further the domain of application may be severely restricted, but we shall insist on a system which will model the task in the "correct" human way. This is a challenge almost totally within episodic culture, with a modest component of mythic culture, but with no component from mimetic culture. The task is, however, far from trivial—and only now beginning to be convincingly solved. The difficulty lies with the conceptual module. Meaning can no longer be reduced to a list or a data base; to succeed we need geometry.

Remark 1 Current philosophy of language distinguishes three aspects of language, syntax, semantics, and pragmatics, which is also the current dogma of applied and computational linguistics (with the necessary addition of a phonological module to deal with speech). This is harmless for simple applications, such as ticket reservation systems, which operate entirely within the phonological-syntactic-semantic range. In more complicated applications, such as dialogue systems, current approaches foresee added pragmatic features—however in an incremental way. We have suggested a different perspective based on a four-stage evolution of culture and cognition, where the stages are cumulative. Granted the correctness of this evolutionary history we would explain the success of some limited existing systems, such as the PONS translation system, by the fact that these systems live almost totally within the fourth theoretical stage. In this perspective, "pragmatics" is not an addition to the syntax-semantics division in order to deal with a number of "rest factors", but a label which masks a number of radically different phenomena. There is a need for a new analysis of "pragmatics" within the broader context of a theory of cognitive development.

8.2 Model Theory and Geometry

In a historical perspective logic and geometry were partners from the beginning, the paradigm being the Euclidean axiomatization of geometry. But today logic and model theory seem to be in a much closer partnership to algebra and arithmetic. The reason for this has both historical and systematical explanations. From Descartes we saw a coordinatization of geometry; in the late 19th century we saw an arithmetization of analysis; Hilbert wanted to prove the consistency of geometry through a reduction to number theory; and Dedekind and Peano gave penetrating logical analyses of the systems of natural and real numbers. At the same time we saw the formalization of logic through the work of Frege, Russell and Whitehead. This line of development culminated in the first order formalization of set theory by Skolem around 1920. Thus everything conspired to give first order logic its prominence.

A model for first order logic is in essence an algebraic structure; and a fruitful partnership between algebra, arithmetic and logic has been established, starting from the Gödel completeness theory. Natural axiomatizations of geometry and topology would seem to use higher order logic. Since such systems lack the compactness property with respect to the class of intended models, time was not ripe for further collaboration between logic and geometry. There was, however, some activity in the first order model theory of elementary geometry, see Tarski (1959). And a first start on a topological model theory was made by Flum and Ziegler (1980). But geometry was basically missing from model theory.

Interestingly, the link between logic and geometry survived in other fields of study. One example is measurement theory. From the 1950s on there developed links between logic and foundational studies in measurement theory. The starting point was quite algebraic in spirit, scales being seen as systems with an ordering relation. But with multi-dimensional scales the geometric content necessarily becomes more prominent. Studies of perception and measurement led to the notion of a perceptual space, where the prime example is the theory of color space; for an account of these developments, see Suppes et al. (1989).

The need for a richer structure on the conceptual component was convincingly argued in the work on mental models by Johnson-Laird (1983). He was able to show how a simple geometrical representation of knowledge combined with the use of symmetry and invariance properties gave better (i.e. psychologically more plausible) models of reasoning than the standard deductions of formal logic. This insight was carried further by Barwise and Etchemendy in a paper from (1991), *Visual Information and Valid Reasoning*. In this paper we see a first presentation of the "hyper-proof" program which is a system for combined visual and logical reasoning. This has sparked a rich development of systems for heterogeneous reasoning. We should also within the same circle of ideas mention the work of Habel (1990) on representation of spatial expressions. He also argued for a dual coding for the processing of spatial expressions, with both a propositional and a depictorial representation. This is closely linked to the theory of mental models and ideas concerning heterogenous reasoning. In all approaches the inspection of "mental models" combined with rule-based deductions are seen as essential ingredients in spatial reasoning.

We remarked above that even within Montague grammar, where the model theory seems to be given a prominent role, one almost never see in an actual analysis any input from geometry. Most of the work can be carried out at the syntactic level of lambda-terms. But linguistic analysis sometimes forces one to pay attention to geometry. This was a point which I argued in an early paper on Montague grammar; see Fenstad (1979). We shall mention some examples. Our first example concerns the meaning of reciprocals; the geometry of "each other" is quite different in the two examples:

The men were hitting each other.

Five Boston pitchers sat alongside each other.

An interesting classification of the geometry of reciprocals, using the apparatus of generalized quantifiers, has been carried through by S. Peters and co-workers; see Dalrymple et al. (1994).

Locative prepositional phrases present another example where geometry is necessary for the semantical analysis; see Colban (1987). We cannot enter into the details of his theory but restrict attention to one simple example. In this case the prepositional phrase has an adjunct reading:

Peter ran to the school.

The semantical analysis proposed by Colban of this sentence refers to a path or trajectory in space-time ending at the school, such that Peter is in the state of running on the curve. Colban's study presents a full fragment, connecting syntax and semantics through an implemented system.

Much of the work that we have reported on in this section can be conveniently brought under the notion of a *conceptual space* introduced by P. Gärdenfors; see the early papers Gärdenfors (1991, 1993, 1994). We conclude this section with a review of this work, showing in particular its connections to theories of natural kinds, to prototype theory (see Rosch 1978) and to cognitive grammar (see Langacker 1987, 1991; Lakoff 1987). A crucial geometrical notion in this work is the concept of convexity; we shall explain why.

A conceptual space S is given by a number of qualitative dimensions, $D_1, \ldots, D_n$; a point in S is a vector $v = (v_1, \ldots, v_n)$ where each v_i is an element of D_i. Perceptual spaces in the sense of measurement theory (see Suppes et al. 1989) are typical examples of conceptual spaces. As a cognitive construct color space has three dimensions, hue, saturation and brightness, where hue is represented by a circle, saturation and brightness are linear.

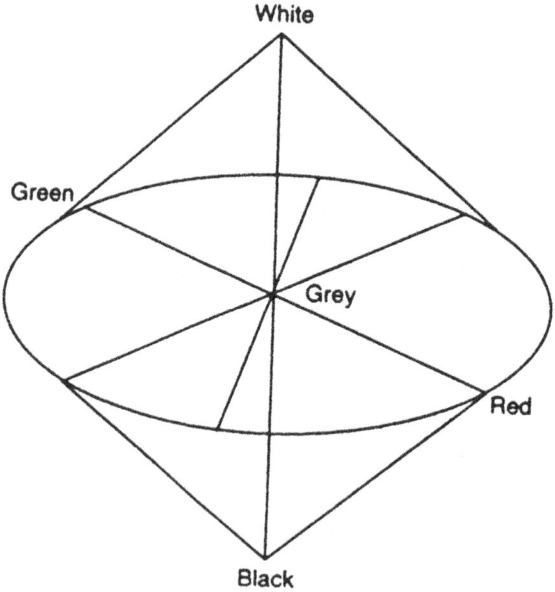

The dimensions determining a given conceptual space may either be inborn or culturally acquired. However this may be, the crucial fact is that each dimension comes equipped with a geometrical or, more generally, topological structure. In color space we have a metric structure defined by measuring length along the appropriate curve (a circle or a straight line). But conceptual spaces can be more abstract than perceptual spaces, one example is the two-dimensional conceptual space generated by the first two formants of vowels frequencies; see Gärdenfors (1991).

P. Gärdenfors has applied the notion of a conceptual space to an analysis of "natural kinds". In traditional model theory a property is interpreted as a subset of the domain. In the theory of conceptual spaces we have access to geometrical notions. A natural property is constructed as a convex subset of the model space. This is obviously the correct way in the case of colors; take any two points in the red sector of the color circle, then any point between is also red, i.e. red as a property is a convex subset of color space. More remarkably, the construction fits the facts in the conceptual space used for the phonetical identification of vowels. How far this assumption is valid, remains to be seen; we shall return to this point in our discussion of geometry and mind.

Here we point out one pleasing application, viz. a valid principle of induction for conceptual spaces: Let P be a natural property in the conceptual space S and let a_1, ..., a_n be points belonging to P. Then the convex hull of a_1, ..., a_n, i.e. the least convex subset of S containing all points $a_1, \ldots, a_n$, is a subset of P; see the discussion in Gärdenfors (1994).

The proposal has also applications to theories of prototypes; see Rosch (1978). In traditional logic properties have been identified with subsets of the domain of interpretation. Granted no further structure there has been no end to the philosophical discussion of what are the "natural" properties and how do we determine when two objects share the same property—when is a tiger a tiger, what exactly are the necessary and sufficient conditions for a chair to be a chair? Prototype theory has developed as an alternative to the logical approach which was based on lists of necessary and sufficient conditions. Natural properties form convex sets in a suitable conceptual space, certain exemplars are more central or typical as examples of the property, i.e. they may serve as prototypes, and the extent of the concept is a convex neighborhood of the accepted prototypes; see Rosch (1978), Mervis and Rosch (1981), Lakoff (1987) and Gärdenfors (1993). Conversely, if we are given a set of categories and have decided on a set of prototypes p_1, ..., p_n for them, then the prototypes determines the categories as a convex partition of the space.

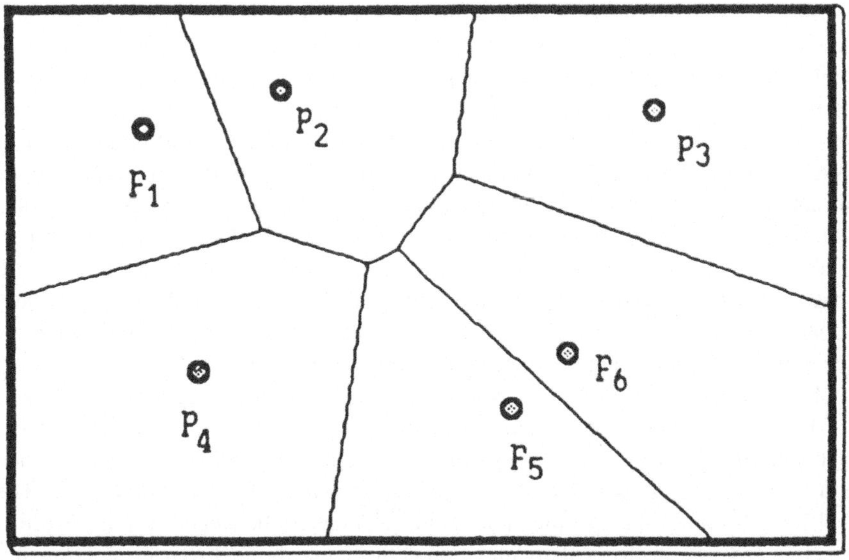

This partition procedure is an example of a Voronoi tessellation; for a comprehensive review of this technique see Okabe et al. (1992). We shall continue this discussion in the next section, translating the discussion into the language of dynamical processes. Here it remains to add a few words on cognitive grammar.

We expressed in the introduction to this chapter some sympathy for the revolt of the cognitive grammarians. But we also noted with some surprise the rather primitive apparatus used by Langacker (1987) in his exposition of the basic ideas. P. Gärdenfors has in "Conceptual Spaces as a Basis for Cognitive Semantics" (see Gärdenfors 1993) shown how model theory with a geometric extension may serve as a basis for cognitive grammar. He agrees that meaning is conceptualization in a cognitive model, not truth conditions in possible worlds. He notes that cognitive models are mainly perceptually determined and that semantical elements are based on spatial or topological objects. He argues in agreement with the cognitive grammarians that semantics is primary to syntax and that concepts defines a "correct" framework for cognitive grammar. We agree—but only up to a point. At this juncture we will only recall that "cognitive space" is the end result of a complex development. This means that conceptual space and linguistic engineering need not live in the same region of cognitive space. However, in order to draw the correct distinctions we need to proceed from grammar and geometry to mind.

8.3 Geometry and Mind

We have so far made a journey from grammar via semantics to geometric structure. In this part we will start out with some remarks on brain dynamics and explain how the associated processes lead to certain geometric structure spaces. The main suggestion of this chapter is the proposal to identify the geometric structure space derived from brain dynamics with the geometric model theory discussed in the last section, the link being the identification of the notion of a "natural property" seen as a convex region of logical model space with the property of being a domain of attraction of an attractor of the brain dynamics; we shall briefly discuss both color space and prototype theory from this point of view to highlight the connection. A theory of mind would then be founded on the theory this class of geometric structure spaces, and granted the identifications mentioned above, we would have a seamless connection from grammar via geometry to mind. Such are the ambitions; we shall see how far they are able to withstand the complexities of the real world.

The affirmative side of the story was well presented in an early review article by J.-P. Changeux and S. Dehaene, *Neural Models of Cognitive Functions* (Changeux and Dehaene 1989). They give a survey of work at the interface between cognitive science and neuroscience, hoping to build bridges rather than enlarge on differences. "The real issue becomes the specification of the relationship between a given cognitive function and a given physical organization of the human brain. From an experimental point of view, our working hypothesis (rather than philosophical commitment) is that levels of organization exist within the brain at which a type-to-type physical identity might be demonstrated with cognitive processes". In their article they review, in particular, work done on short-time and long-time memory and the recognition, production and storage of time sequences. They deliberately pass by a discussion of complex cognitive functions such as problem solving and language processing, since at this stage of development the neurobiological basis is too complex and the relevant data too few for a fruitful mathematical modeling activity. But this is at most a temporary restriction; their final conclusion is that "it is timely to approach cognition in a synthetic manner with the aim to relate a given cognitive function to its corresponding neural organization and activity state". We shall have more to say about the strong assumption of one-to-one correspondence between cognition and anatomy as claimed by Changeux and Dehaene. But first we shall briefly recall another, but related approach to cognitive modeling.

Our next example is the theory of *attractor neural networks* developed by D. J. Amit; see his book *Modeling Brain Function* (Amit 1989). An attractor neural network consists of a fine set N of nodes or neurons, $a_1, \ldots, a_N$ each of which can in the simplest case be in one of two possible states, $s_i = 0$ or $s_i = 1$. Each pair (a_i, a_j) of neurons is connected through a function J_{ij} which measures the influence or the synaptic efficacy which the node a_i may have on the node a_j. We assume that the process is a discrete time process; its dynamics will be given by the two equations:

$$s_i\,(t+1) = ch\,(h_i\,(t+1) - T_i) \tag{8.1}$$

$$h_i\,(t+1) = J_{i1}s_1\,(t) + \cdots + J_{iN}s_N\,(t) \tag{8.2}$$

The system operates as follows: At time t the system is in stage $s = (s_1, \ldots, s_N)$. At each node a_i there is defined a local field h_i. This local field is at stage $t + 1$ determined by the current stage s and the channel weight functions J_{ij} according to Eq. (8.2). The function T_i represents a threshold level at node a_i. If the field h_i at a_i at time $t + 1$ is greater than T_i then the neuron a_i fires, i.e. the value of $s_i(t + 1)$ is set to one; if h_i is less than T_i, then neuron a_i is inactive, i.e. the value of $s_i(t + 1)$ is set to 0. The function ch formalizes this description, i.e. $ch(n)$ is 1 or 0, depending upon whether n is positive or negative; this is the content of Eq. (8.1).

The dynamic behavior of such systems can be quite complicated; we recommend Amit's book as an excellent guide to the field; see also Scott Kelso (1995). The analogy to spin-glass theory in non-linear statistical mechanics is striking. Assuming full connectivity, i.e. J_{ij} is defined for each pair i, j, and symmetry, i.e. the validity of the equations $J_{ij} = J_{ji}$, the full force of mean field theory comes into play and it is possible to analyze the dynamics of an attractor neural network in great detail, in particular, to give a rather full description of the set of attractors of the system. From a neurobiological point of view, the assumption of full connectivity and symmetry is suspicious, if not outright false. But as Amit argues, they form a convenient starting point; through some further "robustness studies" it is also possible to see how the assumptions may be relaxed in order to obtain models more faithful to neurobiological facts.

For the moment let us play with the simplified version of an attractor neural network. It is easy to see how a system defined by the Eqs. (8.1) and (8.2) can be made to memorize certain patterns: In fact, let $p^1, \ldots, p^K$ be K prescribed patterns, i.e. each p^k is an N-tuple of 0's and 1's,

$$p^k = \left(p_1^k, \ldots, p_N^k\right)$$

Define the synaptic connectivity by the equations:

$$J_{ij} = \left(p_i^1 p_j^1 + \cdots + p_i^K p_j^K\right) / N \tag{8.3}$$

Then a simple argument shows that the patterns $p^1, \ldots, p^K$ are the fixed point attractors of the dynamics defined by the Eqs. (8.1), (8.2), and (8.3). With all our simplifying assumptions the system also has an energy function defining a suitable geometric phase space of the system. The set of configurations p^k, being the attractors of the system, determine both the local minimum locations on the energy surface, as well as their domain of attraction. We suggest that this "geometric" approach could be seen as the proper starting point for a phenomenological theory of mind.

In the previous section we saw how the notion of a color space could be reconstructed as a conceptual space. Colors correspond to convex regions in the space; let us for the moment think of hues only, i.e. focus on the color circle. According to E. Rosch colors have prototypical properties, thus the geometry of color space is determined by a Voronoi tessellation based on a finite set of prototypical exemplars. In a similar way color prototypes can be interpreted as a fixed set of patterns to be stored by a suitable attractor network; for a related discussion of prototypes and neural networks see also Chap. 11 of Anderson (1995). In these cases the prototypes are the attractors of the system and the concept of color corresponds to a domain of attraction in the energy surface of the system. This would give a reasonable dynamics for color perception. The correspondence between convex geometry and the dynamics of attractors is quite close; granted sufficient regularity assumptions the claim is that the two accounts tell basically the same story. In this way we see a connection between grammar and mind—the link being geometry.

Remark 2 The store and retrieval of temporal sequences is a first important step beyond the "passive" networks described by Eqs. (8.1) and (8.2) above. D. J. Amit has constructed an attractor neural network that is able (in a precise technical sense) to count chimes; see Amit (1989). Amit sees this as a tentative step into abstract computation. His theory should be reviewed in the context of a theory of cognitive development. Recognizing and reacting to a fixed number of chimes may be an activity of episodic mind only. But theoretical arithmetic, e.g. the recent proof of Fermat's theorem, is beyond any doubts an activity of the fourth stage of cognitive development. It is therefore a questionable exercise to attempt an account of mathematics only in terms of neural networks, i.e. as an activity within an individual mind. The act of doing mathematics is rather a subtly shared activity involving several cognitive stages; see the discussion on the changed role of biological memory in Donald (1993). We shall add some further comments on this Remark in the Postscript to Chap. 9.

Returning to the main text we would like the picture drawn by Amit to be true, and it may, in principle, be so as an account of an individual mind. But what we today know about actual cognitive processes in the brain, tells us that the simple and sometimes one-to-one connection between cognition and neuronal activities postulated by current work in neural network theory is currently far off the mark. Let me briefly point to some of the facts. On the positive side we have the wealth of results obtained through such techniques as PET scanning; see e.g. Kosslyn and Koenig (1992) and Poser and Raichle (1994). This has made possible quite detailed models of cognitive functions, models which are well grounded in anatomical facts. But they also points to the complexity of the link between cognition and anatomy. One striking example is a study of lexical access which is reported in Posner and Raichle (1994); see chapter five of their book. In this chapter they study a hierarchy which starts with the act of passively viewing displayed words, through the gradually more complex tasks of listening, speaking and generating words. Each task in this hierarchical experiment is seen to activate distinct sets of brain areas. The information is precise but true understanding is complex and still far away. I shall only make two comments: PET studies do identify brain areas active in a give cognitive task, but they tell us little about the mechanisms involved. Typically, Kosslyn and Koenig have in their book a chapter on computations in the brain, but little use is made of this in this chapter in later parts of the book—simply because the "true" computational structures are not known; for a review see Churchland and Sejnowski (1992). My second comment is that while there certainly is a correspondence between cognitive function and neural structure and activity states, the correspondence is not necessarily one-to-one; different activity areas with different architectures may generate similar geometries.

Both cases argue for an independent phenomenological theory of mind. This means a study of the geometry with presupposing a detailed knowledge of the underlying dynamic behavior. This is not an unusual situation in science; to mention one example we know that we in equilibrium thermodynamics are—at least in principle—able to reduce the phenomenological theory of heat to molecular motion, but that we in non-equilibrium theory are still largely ignorant of the dynamics and therefore must introduce separate equations for the phenomenological level. A beautiful example of this strategy is found in A. Turing's study of the chemical basis of morphogenesis, see Turing (1952). He assumed the existence of two active chemical substances—so-called concentrations of morphogenesis in his language—which generate a non-linear dynamics governed by a pair of coupled diffusion-reaction equations. Within this model he was able to show how the geometry generated by the process could explain certain morphological phenomena, one example being the process of gastrulation. One reason for mentioning this work is that Turing's model is not only a beautiful example in itself, but that it may also teach a lesson how to model the connection between geometry and mind; for a more recent exposition on pattern formation and diffusion-reaction equations see Meinhardt (1995).

The task of modeling this connection was attempted by Thom around 1970; see his papers *Topologie et Linguistique* (Thom 1970) and *Langage et Catastrophes: Eléments pour une Sémantique Topologique* (Thom 1973). The geometric locus for

Thom is also an "energy surface" which is supposed to be derived from an underlying brain dynamics. Thom, however, does not explicate the dynamics. His discussion proceeds at a purely geometrical or phenomenological level. In the 1970 paper he classifies spatio-temporal verb phrases in terms of singularities in the energy surface, and derives a "natural classification" of such verb phrases in terms of his classification of singularities into seven classes. This should be seen in connection with the classification of "natural kinds" in terms of convex regions and domains of attraction which was discussed above. In the second paper he develops a more systematic discussion. It would be beyond the scope of this chapter to survey this work here. Suffice it to say that he claims, in complete agreement with our previous discussion, that a noun phrase is described by a potential well in the dynamics of mental activities and a verb phrase by an oscillator in the unfolding space of a spatial catastrophe. However, Thom's work has not had the influence on theoretical linguistics that it merits. There seems to have been an incompatibility of minds. Thom severely criticized Chomsky for his combinatorial approach as being totally inadequate as a theory of linguistic meaning. Linguists, unfamiliar with the mathematics, saw little relationship between Thom's "speculations" and their science. One early attempt to bridge the gap can be found in my paper on Montague grammar *Models for Natural Languages* (Fenstad 1979). I argued with reference to Thom's work for a link between him and Chomsky based on a geometrization of the model theory of Montague. But I did not pursue the topic further at that time.

In conclusion I would like to discuss some early work on sentence processing which fits into the picture drawn here. This is work done by G. Kempen and T. Vosse (see Kempen and Vosse 1990). They have developed an attribute-value approach to grammar called segment grammar.

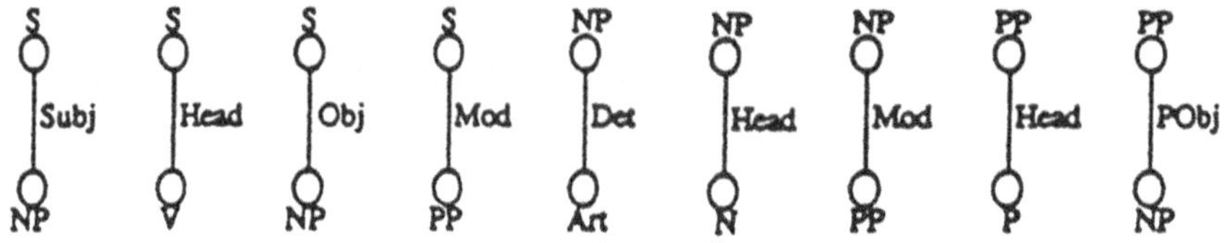

The grammar has two kinds of signs. One type—the so-called syntactic segments exhibited in the figure above—has a purely structural function. The other type—the lexical signs—are signs which carry linguistic meaning; see the following example taken from the same paper.

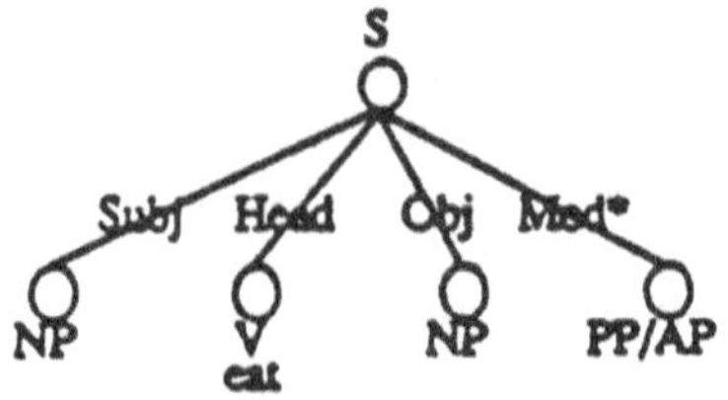

The authors introduces the notion of a unification space, which, to use their words, is a kind of "test tube" filled with a mixture of syntactical and lexical signs. The structural signs are always assumed to be present in the tube; from time to time lexical signs are added and the tube well taken. A reaction occurs and complete sentential signs are crystallized.

The pictorial language has a complete algorithmic interpretation. Signs combine through unification procedure; see Kempen and Vosse (1990).

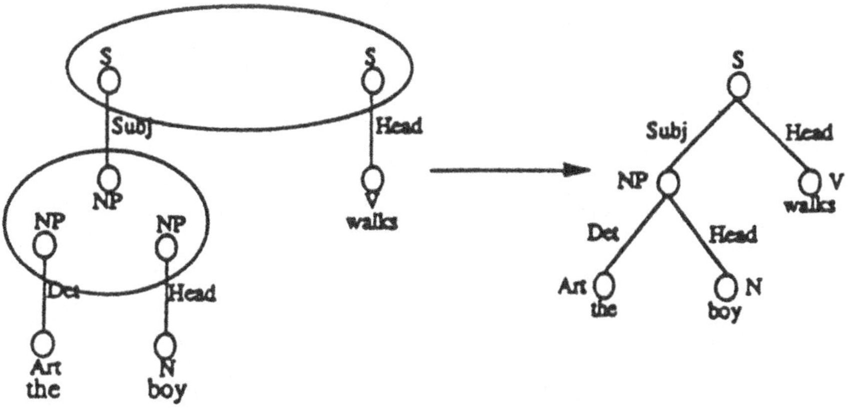

The dynamics behavior of the unification process is defined by a simulated annealing algorithm. Such algorithms, which have their root in statistical mechanics and material science, have become very popular in the field of combinatorial studies. In a sense they are very universal in nature, but not always efficient in execution. Since such algorithms are well known, we shall not give any details but rather suggest another possibility for the dynamics of unification spaces.

In a number of recent works W. J. Freeman and co-workers at Berkeley have developed an alternative approach to neural modeling; for a convenient survey see the papers Skarda and Freeman (1987, 1990). Their particular concern is the olfactory system, but the aim is a general model of brain dynamics. They—of course—recognize a neuronal level of description but claim that it is not the right level for a general theory of brain dynamics. They find a philosophical alibi in the work of the French phenomenologist Merleau-Ponty and describe the dynamics as an example of nonlinear and, hence, self-organizing systems; they could as well have made a reference to the work of Turing (1952). They argue for a chaotic dynamics, in particular, in order to account for the speed in the recognition of smells. In the *Behavioral and Brain Sciences* paper (Skarda and Freeman 1987) there is the usual critical review section. There may be critical comments, but the validity of a description at a phenomenological level cannot be denied. This was also the point we argued above. The complexity of brain mechanics in the execution of cognitive tasks is so intricate that er are in any case forced back to a phenomenological level of description. It remains to be seen whether this is a matter of convenience or of principle—Changeux and Freeman would probably differ on this point.

We have in this chapter traced a line from grammar to mind, the link being geometric model theory. Current computational linguistics concentrate to a large extent on the syntax/semantics interface. In most of this work the meaning component—the conceptual module—is reduced to a data base. This is adequate as long as we restrict attention to the final and theoretical stage of cognitive development, which is characterized by external symbol storage, and which form the base for current developments in language engineering. But it is not adequate as a basis for a theory of linguistic behavior in human beings. This why we have insisted on including all stages of cognitive development in our account in order to appreciate the complexities involved. This perspective led to a geometrization of model theory as a basis for a phenomenological model of mind.

But mind could also be he starting point for our quest for understanding the connection between language and meaning. Our working hypothesis would then be that a phenomenological theory of "mind" is nothing but a branch of geometric model theory. The geometry is the basic object which introduces the lexical items, the rest—grammar, logic and mathematics—are formal tools used in the study of its structure. In particular, it seems possible to take a modified notion of unification space as a starting point for an account of language and speech, but in order to do so we need to explore a wider range of possible dynamics and search for new algorithmic approaches.

References

Anderson, J. A. (1995). *An introduction to neural networks*. Cambridge, MA: MIT Press.

Amit, D. J. (1989). *Modeling brain function*. Cambridge: Cambridge University Press.

Barwise, J., & Etchemendy, J. (1991). Visual information and valid reasoning. In W. Zimmermann & S. Cunningham (Eds.), *Visualization in mathematics*. Washington, D.C.: Mathematical Association America.

Barwise, J., & Perry, J. (1983). *Situations and attitudes*. Cambridge, MA: MIT Press.

Changeux, J. P., & Dehaene, S. (1989). Neuronal models of cognitive functions. *Cognition, 33*, 63–109.

Chomsky, N. (1965). *Aspects of the theory of syntax*. Cambridge, MA: MIT Press.

Churchland, P. C., & Sejnowski, T. J. (1992). *The computational brain*. Cambridge, MA: MIT Press.

Colban, E. (1987). *Prepositional phrases in situation semantics*, Appendix A of Fenstad et al. (1987).

Dalrymple, M., Kanazawa, M., Mchombo S., & Peters, S. (1994). What do reciprocals mean? In *Proceedings of the Fourth Semantics and Linguistics Theory Conference*, Cornell University Working Papers in Linguistics.

Donald, M. (1990). *Origin of the modern mind: Three stages in the evolution of culture and cognition*. Cambridge, MA: Harvard University Press.

Donald, M., et al. (1993). Precis and discussion of origin of the modern mind. *Behavioral and Brain Sciences, 16*, 737–791.

Dyvik, H. (1993). *Exploiting structural similarities in machine translation, institute of linguistics*. Bergen: University of Bergen.

Fenstad, J. E. (1979). Models for natural languages. In J. Hintikka et al. (Eds.), *Essays in mathematical and philosophical logic*. Dordrecht: D. Reidel Publishing Company.

Fenstad, J. E., Halvorsen, P. K., Langholm, T., & van Benthem, J. (1987). *Situations language and logic*. Dordrecht: D. Reidel Publishing Company.
Fenstad, J. E., Langholm, T., & Vestre, E. (1992). Representations and interpretations. In M. Rosner & R. Johnson (Eds.), *Computational linguistics and formal semantics*. Cambridge: Cambridge University Press.
Fenstad, J. E. (1996). Partiality. In J. van Benthem & A. ter Meulen (Eds.), *Handbook of logic and linguistics*. Amsterdam: North-Holland.
Flum, J., & Ziegler, M. (1980). *Topological model theory, Springer lecture notes in mathematics*. Heidelberg: Springer.
Gärdenfors, P. (1991). Framework for properties. In L. Haaparanta et al. (Ed.), *Language, knowledge and intentionality* (Vol. 49). Helsinki: Acta Philosophica Fennica.
Gärdenfors, P. (1993). *Conceptual spaces as a basis for cognitive semantics*, Department of Philosophy. Lund: Lund University.
Gärdenfors, P. (1994). Three levels of inductive inference. In D. Prawitz et al. (Eds.), *Logic, methodology and philosophy of science IX*. Amsterdam: North-Holland.
Habel, C. (1990). Propositional and depictorial representation of spatial knowledge: The case of path-concepts. In R. Studer (Ed.), *Natural language and logic, lecture notes in artificial intelligence*. Heidelberg: Springer.
Johnson-Laird, P. J. (1983). *Mental models*. Cambridge: Cambridge University Press.
Kamp, H., & Reyle, U. (1993). *From discourse to logic*. Dordrecht: D. Reidel Publishing Company.
Kaplan, R., & Bresnan, J. (1982). Lexical-functional grammar. In J. Bresnan (Ed.), *The mental representation of grammatical relations*. Cambridge, MA: MIT Press.
Kempen, G., & Vosse, T. (1990). Incremental syntactic tree formation in human sentence processing. *Connection Sciences, 1*.
Kosslyn, W., & Koenig, O. (1992). *Wet mind*. New York, NY: The Free Press.
Lakoff, G. (1987). *Women, fire and dangerous things*. Chicago: Chicago University Press.
Langacker, R. W. (1987 and 1991). *Foundation of cognitive grammar* (Vols. 1 and 2). Stanford CA: Stanford University Press.
Lönning, J. T. (1989). Mass terms and quantification. *Language and Philosophy, 10*.
Meinhardt, H. (1995). *The algorithmic beauty of sea shells*. Heidelberg: Springer.
Mervis, C., & Rosch, E. (1981). Categorization of natural objects. *Annual Review of Psychology, 32*.
Montague, R. (1974). In R. Thomason (Ed.), *Formal philosophy*. New Haven, CT: Yale University Press.
Okabe, A., Boots, B., & Sugihara, K. (1992). *Spatial tessellations, concepts and applications of Voronoi diagrams*. New York, NY: Wiley.
Pollard, C., & Sag, I. (1987). *Information-based syntax and semantics* (Vol. 1). Stanford, CA: CSLI Lectures Notes.
Popper, K. (1972). *Objective knowledge*. Oxford: Clarendon Press.
Posner, M. I., & Raichle, M. E. (1994). *Images of mind*. New York, NY: Scientific American Library.
Rosch, E. (1978). Prototype classification and logical classification. In E. Scholnik (Ed.), *New trends in cognitive representations; Challenges to Piaget's theory*. Hillsdale, NJ: Lawrence Erlbaum Associates.
Scott Kelso, J. A. (1995). *Dynamic patterns: The self-organization of brain and behavior*. Cambridge, MA: MIT Press.
Skarda, C. A., & Freeman, W. J. (1987). How brain make chaos in order to make sense of the world. *Behavioral and Brain Sciences, 10*.
Skarda, C. A., & Freeman, W. J. (1990). Chaos and the new science of the mind. *Concepts in Neuroscience, 1*.
Suppes, P., Krantz, D. H., Luce, R. D., & Tversky, A. (1989). *Foundations of measurement* (Vol. 2). New York, NY: Academic Press.
Tarski, A. (1959). What is elementary geometry. In L. Henkin et al. (Ed.), *The axiomatic method*. Amsterdam: North-Holland.

Thom, R. (1970). Topologie et Linguistique. In *Essays on topology*. Heidelberg: Springer.

Thom, R. (1973). Langage et catastrophes: Eléments pour une sémantique topologique. In J. Palis (Ed.), *Dynamical systems*. New York, NY: Academic Press.

Turing, A. M. (1952). The chemical basis of morphogenesis. *Philosophical Transactions of the Royal Society of London. Series B, Biological Sciences, 237*(641), 37–72.

Vestre, E. (1987). *Representasjon av direkte spørsmål*. In Cand. Scient. Thesis, Department of Mathematics, University of Oslo.

White, L. A. (1947). The locus of mathematical reality: An anthropological footnote. *Philosophy of Science Journal, 14*(4), 289–303.

Chapter 9
On What There Is—Infinitesimals and the Nature of Numbers

Abstract This essay will be divided into three parts. In the first part we discuss the case of infinitesimals seen as a bridge between the discrete and the continuous. This leads in the second part to a discussion of the nature of numbers. In the last part we follow up with some observations on the obvious applicability of mathematics. The main body of the text is followed by a Postscript with some further thoughts on what numbers are.

Keywords Numbers and cultural foundation · Numbers—evolutionary theory—neuroscience · Cultural invariance of numbers · Abstract computations—physical calculations

9.1 Part A. Infinitesimals

Mathematical intuition is founded upon two sources, the *continuous* and the *discrete*, and the tension between the two has always played a central role in mathematical thought. This is true not only in the discussion of the foundation of mathematics, but also in the use of mathematics in the modeling of natural phenomena.

The intuition of the *discrete* has its roots in counting and labelling; one way of connecting it with the *continuous* is through the act of *measurement*. Measurements create points and intervals on the geometric line, and these points, as well as the length of the intervals, can then be marked or named by numbers. We soon experience that, in addition to the integers and the fractions or rationals, we need an extension of the notion of number to measure e.g. the diagonal of a square. So much for our immediate intuition. The set of reals can in different ways be seen as the completion (in the sense of topology) of the rationals. And up to this day most mathematicians would claim that the "real" points created in this process exhaust the geometric line. But is it necessarily so?

Let us discuss the case of normalized counting measures. If we have a finite set F we can measure the size of any subset A relative to F by counting the number of elements in A and divide by the total number of elements in F. For any natural number

J. E. Fenstad, *Structures and Algorithms*, Logic, Argumentation & Reasoning 15,
https://doi.org/10.1007/978-3-319-72974-9_9

n let F_n be the following set of points in the unit interval I from 0 to 1: $1/n$, $2/n$, …, $1 = n/n$. Using the normalized counting measure on F_n we can measure the size or length of any subset A of F_n. If we increase n we get a better and better approximation to the unit interval I, and hence the associated measures would be better and better approximations to measure length and size of subsets of the interval. But we have a problem, the limit of the sets F_n and the associated counting measure do not exist. Using the set of real numbers as a limit space will not do, we would then have to give each real number an infinitesimal weight. But "most mathematicians" deny that infinitesimals exist.

An infinitesimal is a number x such that $x > 0$, but $x < 1/n$ for all natural numbers n. Such numbers do not exist in the official version of the real numbers **R**. But all through the history of mathematics they have been used in an informal way as a bridge between the discrete and the continuous. Archimedes introduced infinitesimals as a method of discovery in his dissertation *The Method*. This set a style, and the use of infinitesimals was an important heuristic instrument in the discovery of the rules of calculus in the sixteenth and seventeenth century. Leibniz in the late seventeenth century almost got it right, he regarded infinitesimals as an extended number system with the same rules of reckoning as the known numbers. And he effectively used the extended number system to build a foundation for the new "infinitesimal" calculus.

But turning the heuristics of infinitesimals into a correct method was beyond the mathematicians of the time. This led some, one example being Bishop Berkeley in the early eighteenth century, to declare the notion of infinitesimals as inconsistent and not to be part of the new calculus. But as we have learned from the Feynman integral of today, what is inconsistent in a narrow formal sense, may still be a useful tool for insight and explanation. So with the infinitesimals, they survived, chiefly among the "natural philosophers", i.e. the physicists.[1]

A secure logical foundation for analysis emerged only in the 1870s through the work of Cantor, Dedekind and Weierstrass. And the new foundation had no room for infinitesimals. This was strongly expressed by A. Fraenkel in his well known text on Mengenlehre: "Bei dieser Probe hat aber das Unendlichkleine restlos versagt".[2] In a certain sense, set theory and its use in the foundation of analysis marked a victory for the discrete over the continuous: In the beginning there is the empty set—the rest, including the geometric continuum, is an exercise in set theoretic constructions.

The standard view on analysis was firmly seconded by Skolem. In a lecture[3] to a Scandinavian congress in 1929 he argued against using infinitesimals as a foundation:

"We know that such entities can be introduced—they can be exhibited in so-called non-archimedian number systems—but it is not possible to construct a calculus of infinitesimals on such a foundation". There may indeed be entities in the

---

[1] For a detailed study of the history of infinitesimals from Leibniz to Robinson see Katz and Sherry (2013); see also Ehrlich (2006) and Feferman (2009) for a more extensive discussion of possible conceptions of the continuum.

[2] "In this test, however, the infinitely small has completely failed." Fraenkel (1928), p. 116. The test referred to is how to prove a mean value theorem for arbitrary intervals, including infinitesimal ones; see Sect. 6.1 on the Klein-Fraenkel criterion in Kanovei et al. (2013).

[3] Skolem (1929).

mathematical universe that exhibit the "defining" properties of infinitesimals, but the question of interest to us is if infinitesimals exist in an (elementary) extension of the real number system, and if there is enough room on the "real" geometric line to faithfully represent the extended number system. This is what Skolem denies in his 1929 lecture.

But do not always listen to what great men say. In 1933 Skolem constructed his non-standard extension of the natural number system,[4] i.e. he constructed an "object" ***N** as a proper elementary extension of the natural number system **N**, where "elementary extension" means that all the basic rules of arithmetic are preserved in the extension from **N** to ***N**. As a proper extension ***N** must contain at least one infinitely large number, hence the inverse of any infinitely large number in ***N** will be a true infinitesimal. This was observed by Abraham Robinson in 1960.

"In the fall of 1960 it occurred to me that the concepts and methods of contemporary Mathematical Logic are capable of providing a suitable framework for the development of the Differential and Integral Calculus by means of infinitely small and infinitely large numbers."[5]

Starting from the infinitely large numbers and the infinitesimals, Robinson was lead to a richer set of points on the geometric line, usually called the hyper-reals and denoted by ***R.** We recall that ***R** is, in the technical language of logic, a proper (and saturated) elementary extension of **R**, containing both infinitesimals and their inverses, the infinitely large numbers; see in addition to Robinson also Keisler[6] and Fenstad.[7] The construction of the coordinate field ***R** is from our point of view primarily a method of constructing new points on the geometric line. The coordinate set ***R** has strong closure properties and we may, similarly to the standard approach, "forget" the ambient geometric space and choose one version of ***R** as the extended geometric line. But this is not, we would insist, correct. The set of "coordinates" ***R** is not the geometric object, but only one method of "naming" points in the continuum. We are therefore at liberty to create different "point-sets", or more correctly, different coordinate structures, on the geometric line for different purposes.[8] We shall later see how this leads to richer possibilities in modeling natural phenomena.

---

[4]Skolem (1934).

[5]Robinson (1966).

[6]Keisler (1994).

[7]Fenstad (1985, 1987).

[8]At this point the reader is advised to turn to Keisler (1994). We have made a virtue of the non-uniqueness of the hyperreal line; others would regard this as an argument against the non-standard theory. It is a fact that the existence and the uniqueness of the hyper-reals depend on the underlying set theory and how these numbers are constructed relative to the chosen set theoretic basis, see Keisler (1994), Sects. 11 and 12. For many, perhaps most, application in the natural sciences the important fact is that infinitesimals exist in the geometric continuum. We should also note that there is an alternative approach to an extended arithmetic continuum, the so-called surreal numbers introduced by J. Conway in 1976; see the extensive discussion in P. Ehrlich (2012), where, in particular, the relationship between Robinson's non-standard reals and Conway's surreals is analyzed.

Let us return to the quote from Robinson. We should note the careful language used by Robinson in this text. We have a "framework", but what is the real content, in what sense do infinitesimals exist? This is Robinson's view in 1966:

"Whatever our outlook and in spite of Leibniz' position, it appears to us today that the infinitely small and the infinitely large numbers of a nonstandard model of analysis are neither more nor less real than, for example, the standard real numbers. This is obvious if we introduce such numbers axiomatically; while in the genetic approach both standard irrational numbers and nonstandard numbers are introduced by certain infinitary processes. This remark is equally true if we approach the problem from the point of view of the empirical scientist. For all measurements are recorded in terms of integers and rational numbers, and if our theoretical framework goes beyond these, then there is no compelling reason why we should stay within an Archimedian number system".[9]

For Robinson, who on several occasions expressed a strong formalist conviction, this is the end of the story: "From a formalist point of view we may look at our theory syntactically and may consider that what we have done is to introduce new deductive structures rather than new mathematical entities."[10]

Not everyone will agree with this point of view. Gödel in some remarks[11] following a talk by Robinson at the Institute for Advanced Study in March 1973 seems to maintain that there is no ontological difference between the integers, the rationals, the irrationals, and the infinitesimals. By a series of "quite natural steps" we may become familiar with and gain insight into what exists.

***R** can also be put to useful work in modeling natural phenomena. First let us return to the counting measure discussed above; let now H be a hyper-finite natural number, divide the unit interval into H points, give each point the infinitesimal weight 1/H, then the so-called standard part of the hyper-finite counting measure gives us the "correct" way of measuring length and size with respect to the unit interval; we have, in fact, constructed the Lebesgue measure. All important properties of this measure follow by transfer from the corresponding properties of finite normalized counting measures. We could continue in this way: Hyper-finite models of statistical mechanics is a direct and "geometric" way to obtain the thermodynamic limit; hyper-finite models of quantum field theory allow us to define the field directly in a point-wise manner.[12] In all these cases we have seen how infinitesimals and the infinitely large have been used to handle singularities and infinities occurring in the natural sciences. Note that singularities and infinities occur within the domain of finite measurements and are so labeled because the standard framework has no direct way of dealing with them. Nonstandard analysis gives us a direct and geometric method for the "taming" of such infinities.

Robinson was careful to acknowledge his debts to Skolem's work on non-standard models of arithmetic. Skolem died in 1963 and it is highly doubtful if he ever learned

---

[9]See footnote 5.

[10]Robinson (1965).

[11]Gödel (1974).

[12]Albeverio et al. (1986), Chap. 7.

of the new infinitesimal analysis. We cannot therefore know if he would have changed his view from 1929 and admitted that there is room for infinitesimals on the geometric line. But, with reference to the Robinson-Gödel debate, we believe that the case of infinitesimals is important for a proper understanding of what there is and how mathematical knowledge is possible. Robinson's position is an example of the view that mathematical knowledge is not knowledge of abstract mathematical objects. In contrast, Gödel seems to believe in the necessity of the existence of such objects and that mathematical knowledge thus is knowledge about these objects. We agree with Gödel, infinitesimals exist as firmly as the integers, the rationals and the irrationals do.[13] But the "next natural step" presupposes a first step. What is the nature of numbers, in what sense do they exist?

9.2 Part B. The Nature of Numbers

"Es kommt doch auch bei der Begründung der Mathematik auf die Sache und nicht auf die Bezeichnung an". The words are Skolem's and occur towards the end of his 1923 paper on the *Begründung der elementären Arithmetik durch die rekurrierende Denkweise*.[14] Thoralf Skolem, who wrote this paper in 1919, had at that time become acquainted with the approach to elementary arithmetic in the *Principia Mathematica* of Russell and Whitehead, and he was not at all convinced. Natural numbers and their fundamental properties need not—and should not—be justified through an exercise in abstract logic. This seems to be a strongly held belief of Skolem, firmly rooted in his early education and work experiences. By 1919 he had already been exposed to a wide range of problems and projects in both pure and applied mathematics. His master thesis in mathematics from 1913 was a profound study of the algebraic logic of Peirce and Schröder. His first employment was as a student assistant to the well-known Norwegian physicist K. Birkeland in the years 1909–1914. His special assignment was to help with the mathematics of several geophysical phenomena. In 1917 he published his first major mathematical paper *Untersuchungen über einige Klassen kombinatorischer Probleme*.[15] It is therefore not surprising that he was not at all convinced when he in Göttingen in the winter 1915–1916 first learned about the Russell and Whitehead approach to elementary

[13]Further information about the theory and the applications of non-standard analysis can be found in Albeverio et al. (1986), Keisler (1994) and the recent volume edited by van den Berg and Neves (2007). For our purposes we recommend the contribution by Keisler, *The Strength of Nonstandard Analysis*, Keisler (2007). The debate between Gödel and Robinson is discussed in two "introductory notes" in Gödel's Collected Works; see Fenstad (1990) on Gödel's remarks at the Institute in 1973, and Machover (2003) on the Gödel-Robinson correspondence.

[14]Skolem (1923). The German text is translated as "Yet, even in providing a foundation for mathematics it is the substance that is important, not the notation" in van Heijenoort (1967), p. 333. I am not happy with the translation 'substance' of the word 'Sache'; I would argue that 'Sache' in this context refers to what there is.

[15]Skolem (1917).

arithmetic in the *Principia Mathematica*. We get the impression that Skolem through his work in algebra, combinatorics and geophysics "knew" what numbers are, and that his 1923 paper can be seen as a protest against the the approach of Russell and Whitehead. In this paper he takes the first steps towards a more "correct" approach by developing what has later become known as primitive recursive arithmetic, and we would argue that to choose the "rekurrirende Denkweise" as a foundation for "real" mathematics was almost to be expected from an expert in combinatorial analysis. He develops his theory within a free variable logic formalism, but he is not entirely satisfied. Too much of "die Bezeichnung" of Russell and Whitehead remains, and he promises a further paper on the real "Sache", which will strictly follow the views of Kronecker that "eine mathematische Bestimmung dann und nur dann eine wirkliche Bestimmung ist, wenn sie mit Hilfe endlich vieler Versuche zum Ziele führt".[16] But the promised paper never appeared.

Neither Kronecker nor Skolem tell us what a number exactly is. But an answer to this problem is, however, only part of the challenge. The world of mathematics contains, as we noted above, in addition to the "discrete" objects of arithmetic, also the "continuous" objects of geometry. Is Russell's logic and Zermelo's axioms for Cantorian set theory an answer to the question of existence? Neither Kronecker nor Skolem would agree. Mathematics is more than an axiomatic exercise in logic and abstract set theory. In a popular lecture given at the Christian Michelsen Institute in Bergen in 1932 Skolem says: "If one works within a completely formalized mathematics, based on a finite number of precisely stated axioms, there is nothing to discuss but questions of consistency and ease of manipulation. But in ordinary mathematical practice, e.g. in the usual studies on continua which are never given by a set of specified formal construction principles, the axiom of choice is, in my opinion, definitely undesirable – a kind of scientific fraud".[17] Leaving the question of "scientific fraud" aside, a challenge to "ordinary mathematical practice" remains, to explain what kind of objects numbers and geometric objects are.

The ontological status of "real" numbers and continua leads on to controversial grounds. Few, if any, would deny that our "world" allows for both physical objects and states of consciousness. Some would further argue for a reduction of the mental to the physical, others would maintain a strict dualism between the two. But a "world" consisting of only material and/or mental objects provides, as much of contemporary philosophy of mathematics bears witness to, a rather awkward framework for an understanding of what numbers are. A third kind of objects seems to be necessary to give a satisfactory explanation of mathematical existence.

---

[16]This is translated as "… a mathematical definition is a genuine definition if and only if it leads to the goal by means of a finite number of trials" in van Heijenoort (1967), p. 333.

[17]The original text is in Norwegian; see Skolem (1933): "Vil man gå over til en konsekvent formalistisk matematikk, basert på et endelig antall eksakt formulerte aksiomer, så er der intet annet å innvende, enn at der kan være spørsmål om motsigelsesfrihet og hensiktsmessighet o. l. Men innenfor den matematiske praksis, slik den vanligvis drives med kontinua, som slett ikke er gitt i kraft av de og de bestemte oppregnede konstruksjonsprinsipper, er utvalgsaksiomet efter min mening avgjort en uting – en slags videnskabelig svindel".

The question of the existence of "ideal" objects has a long history in philosophy and cannot be reviewed here. We shall make use of Popper's theory of objective knowledge as a convenient entering point for our few remarks; see, in particular, the lecture from 1968 *On the Theory of the Objective Mind*. His philosophy, with roots both in Frege and Husserl, recognizes at least three ontologically distinct domains:

"or, as I shall say, there are three worlds: the first is the physical world or the world of physical states; the second is the mental world or the world of mental states; the third is the world of intelligible entities, or of *ideas in the objective sense*; it is the world of possible objects of thought: the world of theories in themselves, and their logical relations; of arguments in themselves; and of problem situations in themselves."[18]

Even if the so-called World 3 is a "man-made product", Popper argues at great length for the objective nature of what it contains, and he highlights, in particular, the causal interaction between this world and the other two. His discussion ranges over a number of themes, among them the possibility of how a World 3 theory could provide an epistemology for the humanistic sciences. But rather than continue with Popper's World 3, I shall briefly turn my attention from philosophy to anthropology.

In a paper[19] published some years ago. I pointed to a striking similarity between the Popperian ontology and the views on mathematical existence expressed by the American anthropologist L. A. White in an essay, *The Locus of Mathematical Reality: An Anthropological Footnote*.[20] White opens his essay by asking: Do mathematical truths reside in the external world, there to be discovered by man, or are they man-made inventions? Does mathematical reality have an existence and validity independent of the human species or is it merely a function of the human nervous system?

White provides an answer to these questions by adding to the two existential categories of *matter* and (individual) *mind* the category of the *mind of the species*. The latter is a term sometimes used by anthropologists in their general discussion of *human culture*.[21] For our purposes we may note the close relationship of this concept to the Popperian notion of World 3. And many of the arguments used by Popper to assert the independent existence of World 3 are similar to the arguments that the anthropologists use to assert the independence of human culture from both the physical world and the individual mind. It is in this notion of culture that our anthropologist sees a solution to the problem of mathematical existence—in his words, there is no mystery about mathematical reality, he notes that mathematics is a kind of primate behavior as languages, musical systems and the penal code are, that mathematical concepts are man-made just as ethical values, traffic rules, and bird cages are man-made:

---

[18]Popper (1972), p. 154.

[19]Fenstad (1985).

[20]White (1947).

[21]If "mind of the species" is too esoteric, substitute the phrase "cultural heritage", see Grattan-Guinness (2004).

"Mathematics in its entirety, its "truths" and its "realities" is part of human *culture*, nothing more ...".[22]

"Mathematical truths exist in the cultural tradition into which the individual is born, and so enters his mind from the outside. But apart from cultural tradition, mathematical concepts have neither existence nor meaning, and of course, cultural traditions have no existence apart from the species. Mathematical realities thus have an existence independent of the individual mind, but are wholly dependent upon the mind of the species".[23]

Our anthropologist is not an isolated phenomenon. The cultural view on the origin and growth of mathematics has a significant support within the mathematical community, one early voice being J. von Neumann.[24] A classical reference is the book "The Mathematical Experience" by P. J. Davis and Reuben Hersh,[25] see also Hersh.[26] A recent voice in support is David Mumford,[27] see, in particular, the section on "The taxonomy of mathematics". A related view is strongly supported by Michael Atiyah in a panel discussion on the interface between mathematics and physics.[28] For a more carefully argued account of mathematics as a "social construct", see the discussion in Cole.[29] We should also mention the earlier contribution of Kitcher.[30] Philosophers of mathematics have not always been convinced by this approach, see the critical analysis in M. Leng.[31]

The arguments of the anthropologist may well convince us as long as the discussion proceeds on this rather general level. But we need, partially in agreement with the philosophers, to understand with much greater precision how matter, mind and culture interact, how mathematical objects come into existence, and how they can be known by man. Some would at this point turn back to philosophy in the tradition of Frege,[32] others would turn to the phenomenology of Husserl, which has currently attracted great interest, a noteworthy example is Gödel's recently published text[33] from 1961. A useful introduction to Husserl's philosophy of mathematics is the survey article[34] written by Richard Tieszen for the *Cambridge Companion to Husserl*. For a post-Gödel development of these ideas see Tieszen.[35] Not being a Husserl

[22]White (1947).

[23]White (1947).

[24]von Neumann (1947).

[25]Davies and Hersh (1980).

[26]Hersh (1997).

[27]Mumford (2000).

[28]Atiyah et al. (2006).

[29]Cole (2008, 2012).

[30]Kitcher (1984).

[31]Leng (2007), pp. 7–15.

[32]Linnebo (2008, 2009).

[33]Gödel (1961). For a general discussion of the relationship between Husserl and Gödel see Føllesdal (1995a, b) and Hauser (2006).

[34]Tieszen (1995).

[35]Tieszen (2011).

scholar I shall refrain from going into the details of his theory of *intentionality* and how it can be used both to elucidate the nature of mathematical objects and to explore how we obtain knowledge of them.

One noteworthy aspect of Tieszen's survey is his sharp criticism of much of current philosophy of mathematics from a Husserlian perspective. In many respects this criticism is similar to the one offered by our anthropologist. But there is one crucial difference, Husserl, according to Tieszen, holds that "mathematical objects cannot be assimilated to social or cultural objects, because social and cultural objects are bound to times and places".[36]

But is this necessarily so? How could the cultural view be strengthen to meet the criticisms of Husserl? This is a question I would like to explore further. We agree that numbers are abstract objects, but how are they related to brain and mind? I shall approach this question by first discussing the evolution of language and how syntactic categories such as noun phrases NP and verb phrases VP became objects of the collective mind. This is a topic which I have discussed at greater length in the book *Grammar, Geometry and Brain*,[37] where I used as a framework the evolutionary theory of M. Donald on how culture developed.[38]

"According to Donald we can recognize four stages in the evolution of culture and cognition: *episodic culture*, *mimetic culture*, *mythic culture*, and *theoretical culture*. The episodic culture was characterized by the ability to react to a fixed situation in which the individual was placed. ... The transition to mimetic culture extended the limited mode of understanding found in the episodic culture. The individual was now able to decompose the meaning of a specific situation into components and to recombine these in order to develop effective patterns of reaction in other, but similar situations. ... the transition from mimetic to mythic culture was above all characterized by the acquisition of language. ... In the third transition, from mythic to theoretical culture, there are no anatomical changes. We see a transition to a stage where we have access to external symbolic storage."[39]

As noted in Fenstad[40] there is an obvious similarity of the theoretical stage as recognized by Donald to the World 3 introduced by Popper in 1972 and to the "collective mind" in the story told by White in 1947. For present purposes the main interest lies in the transition from the mimetic to the mythic stage, above all in the transition from non-syntactic to syntactic communication. The reader is referred to chapter three of *Grammar, Geometry and Brain* where we discuss some interesting results by Nowak and coworkers[41] relevant to this topic. This work is based on a general model of evolutionary dynamics and shows how an elementary form of "proto-syntax", where a whole message is decomposed into an NP + VP pattern, can be the outcome of an evolutionary process. This decomposition can also be

---

[36]Tieszen (1995), p. 448.

[37]Fenstad (2010).

[38]Donald (1990, 1993).

[39]Fenstad (2010), p. 39.

[40]Fenstad (2010).

[41]Nowak et al. (2000).

seen as a starting point for the iterative property characteristic of human languages. The evolutionary process explains how an NP + VP structure which first occurs in the mind of individuals, evolves into an object in the collective mind of the speech community. And since the structure of the world which lies at the base of the Nowak model, is simple and almost universally shared, we also come to understand how syntactic categories will, as many grammarians strongly insist, be the same across the many existing and seemingly different speech communities, and thus become "true" objects in the collective mind of the human species.[42] The same cultural independence does not seem to be the case with word-order.[43]

We next turn to numbers. There has been much current interest among neuroscientists on how mathematics is related to brain structure. A useful introduction to recent research can be found in the book *The Number Sense. How the Mind Create Mathematics*[44] by Dehaene. His starting point is clear: "Our abstract mathematical constructions originate in the coherent activity of our cerebral circuits and of the millions of other brains preceding us that helped shape and select our current mathematical tools". The author thus clearly recognizes that both the mind of the individual and the collective mind of the species are necessary for an understanding of the nature of mathematics. Part of his work, therefore, reviews brain mechanisms that are believed to underlie mathematical activities. This is no easy task. Experimental techniques can tell us the location of cognitive activities in the brain, but rarely reveal the exact mechanisms, i.e. the equations, that govern these activities. Sometimes neural network simulations can be used to suggest the actual mechanisms, one interesting example is the so-called "accumulator model" proposed by Dehaene and Changeux[45] to simulate how animals extract numbers from their environment. The constructed network can be used to understand rudimentary number processing by man and animals. The model predicts what can be called "number-detecting neurons". And, indeed, there are some interesting indications that such neurons may actually exist.[46] Another early contribution is the book *Modeling Brain Function*[47] by D. J. Amit. In Sect. 5.4 of this book, he constructs a network for counting chimes and is able to write down the exact equation determining its dynamics. In discussion the cognitive relevance of the proposed model he notes: "Once again, one should recall the methodological disclaimer. We do not, by any stretch of the imagination, purport to describe children by the little model counting chimes. Yet it is of interest to observe that a number of features, which one usually associates with human cognitive psychology, do find a simple and well defined echo in this version of a

[42]Fenstad (2010).

[43]Dunn et al. (2011). Recently we have also seen several examples of an experimentalist approach to the origin and development of language, for an introduction see Normile (2012). This is research at an early stage, but it seems to strengthen an evolutionary/cultural point of view. See also the recent analysis in B. Bergen (2012).

[44]Dehaene (1997).

[45]Dehaene and Changeux (1993).

[46]Dehaene (1997), p. 33.

[47]Amit (1989).

neural network".[48] Much have happened since Amit and Dehaene; for a current update concerning our understanding of the cognitive basis of mathematical knowledge, see Cappelletti and Giardino[49] and the special issue on space, time and number of the journal *Trends in Cognitive Science*, in particular, the article by Dehaene and Brannon.[50]

But is this sufficient for an invariant foundation? Will an evolutionary perspective also be needed? In the same way as the Nowak model suggested an objective basis for some syntactic categories, there could be a similar story to tell concerning numbers and other mathematical objects. The Nowak model presupposed a simple "world structure" consisting of individuals and events as an input to the evolutionary process. In analogy to this I suggest that some elementary skills and intuitions about counting and measurement acted as a necessary input to a similar development of mathematics. This does not exclude that some rudimentary part of numerical competence is innate. This seems to be the conclusion in Cappelletti and Giardino: "We have addressed the issue of whether numerical knowledge represents a domain-specific, innate semantic category and we supported this view within the framework of cognitive science by presenting evidence from experimental, developmental and animal psychology, from neuropsychology and neuroimaging studies."[51] But, based on what we know today neuroscience is not necessarily the whole story. We need to explored, as indicated above, the history of how culture evolved: First, a survey of the non-European roots of mathematics[52] reveals a long and complicated history of how numbers and systems of notation developed out of "bones, strings and standing stones". Next, we note the lesson learned from the many interesting and relevant studies of numerical cognition in "primitive" societies today.[53] There are similar results to be told about the emergence of geometry.[54] We further note that a considerable body of "trade and marketplace" mathematical skills, independent of abstract mathematical knowledge, survived for a long time in our culture,[55] and this need also to be part of the story. But mathematics as known today did emerge. It is, as noted above, a difficult and still unresolved question what is ab initio hard wired in the brain and what is the result of cultural evolution and environmental factors. The relationship between culture and the human genome is complicated,[56] but it is definitively a two-way and highly interactive process. There is still much to learn and many sciences will be called upon

---

[48]Amit (1989), pp. 252–253.

[49]Cappelletti and Giardino (2007).

[50]Dehaene and Brannon (2010).

[51]Cappelletti and Giardino (2007), p. 82.

[52]Joseph (1991).

[53]Gordon (2004), Beller and Bender (2008). Gordon concludes his study of the Piraha tribe in the Amazon Basin that what is innate is at most the ability to see specific numbers up to three, the rest is culturally determined.

[54]Suppes (2011).

[55]Ferguson (1993).

[56]Some references here are Laland et al., *How culture shaped the human genome: bringing genetics and the human sciences together* (2010) and Fisher and Ridley, *Culture, genes and the human revolution* (2013).

to help explain the matter, but I would argue that enough is now known to assert that *number* and some basic *syntactic categories* did emerge as invariants across human cultures, hence became true objects of the collective mind of the species. Thus, with reference to Donald,[57] I suggest that early human experiences at the mimetic stage of culture could give a foundation for the later development of language and mathematics, a development which for the individual mind was a characteristic feature of the mythic stage, and for the collective mind of the theoretical stage.[58] And as soon as we have numbers, counting/addition and measurements/multiplication the rest of the Skolem "rekurrirende Denkweise"[59] will follow, a claim I shall return to in part C below.

In arguing for this opinion I differ from Husserl's view on mathematics and culture as quoted above. I agree that social and cultural objects in general are bound to times and places, but in the special cases of numbers and some basic syntactic categories I have argued, using our current understanding of evolutionary theory and neuroscience, that they are invariant, hence true objects of the collective mind of the species. It is in this sense that I would agree with our anthropologist about the nature of numbers and other mathematical objects. This is as far as I am currently prepared to go. Others will have to argue for a possible and similar invariance over times and places of such items in the anthropologist's list as "musical systems, penal codes and ethical values".

Much should still be said about how the brain processes mathematics, and about the complex relationship between brain, mind and culture. And many sciences must contribute to this story. I have pointed to some facts from evolutionary theory and neuroscience, others will stress the importance of a psychological theory of learning in relating the individual learning to the gradual acquisition of knowledge in a culture.[60] There is also much to be learned from current studies of perception,

[57]Donald (1990).

[58]Not everyone will agree to this cultural invariance. It has been a widely shared opinion that the absoluteness of simple arithmetical facts, such as $5 + 7 = 12$, is a strong argument against the cultural dependence of mathematics: If mathematics is culture-dependent, there could be cultures where the meaning of terms are the same, but at the same time $5 + 7$ and 12 are not equal. If this is not possible, the "fact" that $5 + 7 = 12$ would necessarily be culture- and mind-independent. Some would see this as an argument in favor of some form of platonism; see Tieszen (2010). I do not agree with this line of reasoning. I shall argue in the next section that structure is the basic entity, syntax is derived and adapted. Thus if structure is given and a suitable structure-dependent syntax chosen, the arithmetical "facts" will follow with necessity. Therefore, if I say that $5 + 7$ equals 12 and you say no, we have to go back to basic structures. I have argued above for the cultural invariance of numbers and of counting/addition based on what we currently know about neuroscience and anthropology. This is an observed "fact", and it follows that I do not expect to be contradicted when I assert that $5 + 7$ equals 12. I do not deny that there could be cultures, past or future, where facts and basic structures are otherwise. But for cultures in our invariance class numbers are true objects in the "mind of the species", and elementary arithmetical facts are absolute. This is what is needed for the applicability of mathematics.

[59]Skolem (1923).

[60]Suppes (2011).

a suitable reference for mathematicians is the survey[61] by David Mumford. But in addition to "what there is" we also need to have an answer to the question "how do we know". We have in our remarks on Husserl pointed to, but not discussed, the problem how we obtain knowledge of abstract, perhaps, even infinite mathematical objects. We have in our remarks on neuroscience and mathematics told a story of how the brain processes mathematics. But neither Husserl[62] nor neuroscience tell us what the allowable methods of proof and construction are. Thus more is needed on "how we know". We can, perhaps, move closer to an answer to this question when we discuss the applicability of mathematics.

9.3 Part C. The Obvious Applicability

The great Hungarian-American physicist Eugene Wigner is well-known for his deep and difficult applications of mathematics to physics, in particular, of his use of group representation theory to quantum physics. He knew his mathematical tools well, and we would think that he also knew why they were so appropriate. But in a lecture in 1959, *The Unreasonable Effectiveness of Mathematics in the Natural Sciences*, he insists on the following points:

"The first point is that mathematical concepts turn up in entirely unexpected connections. Moreover, they often permit an unexpectedly close and accurate description of the phenomena in these connections. Secondly, just because of this circumstance, and because we do not understand the reason of their usefulness, we cannot know whether a theory formulated in terms of mathematical concepts is uniquely appropriate. ... the enormous usefulness of mathematics in the natural sciences is something bordering on the mysterious ... there is no rational explanation for it."[63]

Against Wigner's "unreasonable effectiveness" we shall argue for an "obvious applicability"; we believe that it is possible to understand how we know. Let us seek a first clue from studies on the origin of language. In a previous section we mentioned an evolutionary model for how syntactic communication evolved. The starting point was a structural awareness or understanding of an environment consisting of both individuals/objects and properties/events. The result of the proposed evolutionary process was a transition from the initial communication in whole messages to a language showing some proto-syntactic form. And no one would question that this language was quite appropriate for what was seen or experienced in the environment.

Our next example is geometry. P. Suppes has in a recent paper, "Why the effectiveness of mathematics in the natural sciences is not surprising",[64] discussed the emergence of geometry. He sees an early beginning in the perceptual processes necessary for humans to survive. He further notes in his paper how human culture

[61] Mumford (2002).

[62] Tieszen (1995).

[63] Wigner (1960).

[64] Suppes (2011).

developed the skill to represent "the structure of the external world in remarkable paintings and drawings", and concludes that the key to the early development of geometry lies in the gradually emerging "structural isomorphism between, in one case a perception and an object or process in the world, and in another, between a mental image and an 'abstract' structure". Modern geometry is thus a complex product of nature, mind and culture, and as such is a "product" obviously applicable to what is seen or experienced in the human environment.

There is a long and complicated story to tell. Measurements and constructions relied on some structural understanding of what we "see", and a "proto-mathematics" developed as a tool to make use of these structures, in particular, in the building and construction trades. It is interesting to note that an implicit knowledge of geometric structures survived up through and well beyond the middle ages within these trades; the gothic cathedrals being an outstanding example.[65] We should also include Leonardo da Vinci in our list of early examples.[66] He developed, both as an engineer and as an artist, a keen sense of structure applicable both to man and nature. And his structures were not frozen in time, in some of his drawings we see a beginning of a geometric transformation theory.[67] But Leonardo did not have the necessary mathematics, his syntax was in a certain sense his "miraculous left hand"—"he grasped connections using nothing more than his pencil that it took other scientists many generations later to formulate in equations".[68] For an instructive example see Eloy,[69] where he "formulates in equations" Leonardo's remarkable observation that the total cross-section of a tree remains the same along its height every time it branches.

We see in these examples that structure is prior to syntax.[70] This is also an approach characteristic of much of logic today. The model structure comes first, "we are primarily interested in logics where the class of structures are those where some important mathematical property is built in, and where the language gives us a convenient way of formalizing the mathematician's talk about the property".[71] A pioneer in the model-theoretic approach was Skolem. His work on the Löwenheim-Skolem theorem was a

---

[65] Ferguson (1993).

[66] See Chap. 2.

[67] Capra (2007).

[68] Klein (2010), p. 181.

[69] Eloy (2011).

[70] Note that I use the word syntax in a rather extended sense, I could also have spoken of structure and representation, or perhaps of objects and tools. Note further that in the case of Leonardo, the "syntax" is his left hand, so our representations can in some cases be physical objects. The main thing is that structure is what we "see" and syntax is what we introduce to make sense of what we "see". The reader may at this point also want to consult Mumford (2002) on perception and pattern theory. The reader will further see a close relationship of our extended use of structure to the so-called prototype theory, which is to understand structure as a kind of "fixed point" for what is seen, see Rosch (1983) and Gärdenfors (2000).
Let me also add one further note of warning, my use of the word "structure" is similar to the informal use of the same word in contemporary applied mathematics and is related, but does not correspond exactly, to what is known as "mathematical structuralism" in the philosophy of mathematics; for a careful discussion of the latter, see Nodelman and Zalta (2016).

[71] Barwise (1985), p. 4.

main step in the development of model theory.[72] And Skolem almost always worked in a model theoretic mode, the structures were always for him the most important. We see this in the way he introduced his well-known Skolem normal form, working with satisfiability, not with provability. We also see it in his introduction of what is now called Skolem functions and in his work on the decision problem of logic. He could use the axiomatic method as a tool, but never accepted axioms as a foundation. For him, as we have argued above, "real" structures were the essence; for this point see also P. Cohen: "So, let me say that I will ascribe to Skolem a view, not explicit stated by him, that there is a reality to mathematics, but axioms cannot describe it."[73]

The art and science of *mathematical modeling* will be our bridge from what we see to how we know. It consists typically of three stages: First an analysis of a given scientific or technological problem leading up to a well-defined *structure* (notice that we in this one sentence have hidden all we have to say about *observations* and *experiments*); next comes *syntax* and "equations", which are used to analyze the properties of the structure; finally the development of *algorithms*, which in a scientific context can be used to make specific predictions and to determine its long time development, and which in a technological context is used to produce "blueprints" for construction and production. But equations normally have their limitations, they cannot always be exactly solved. This adds a further ingredient to the practice of mathematical modeling, *simulations* and graphic displays.[74]

We insist that structure is primary and that syntax is secondary and should be chosen to fit the problems at hand. We should also note that the last step, i.e. the development of algorithms, is not always part of the analysis. While algorithms are important in engineering and applied science, including computational linguistics, there are many examples of fundamental mathematical models in the natural sciences without an immediate algorithmic content.

We need to add a few words on the ancient art and practice of algorithms. Again we meet the name of Skolem, his 1923 paper on primitive recursive arithmetic was important for the development of the general theory of algorithms. His approach seem in general to have been very algorithmic in spirit. Indeed, we get the impression that structures and algorithms are more basic than axioms and proofs in the understanding of mathematics.[75] We note that this is a view in complete agreement with the analysis in Cohen.[76] The general notion of algorithm is in the main due to A. M. Turing; we recommend the collection *The Universal Turing Machine. A Half-Century Survey*,[77] in particular the contribution by R. Gandy.[78] From Turing's theoretical analysis and

[72] Fenstad and Wang (2009).

[73] Cohen (2005), p. 2417.

[74] The reader who wishes a deeper insight into the current praxis of mathematical modeling and large scale scientific computing is strongly advised to consult the recent survey by E. Tadmor (2012).

[75] Skolem (1928), see also the discussion on his attitude to Gödel's completeness theorem for first order logic in Fenstad and Wang (2009), Sect. 4.3.2.

[76] Cohen (2005), p. 2409.

[77] Herken (1988).

[78] Gandy (1988).

the early development of digital computers we have seen the rapid development of large scale scientific computing, which is now an essential part of the art and science of mathematical modeling. This is a story well known and need not be retold here. We shall only make the additional observation that Turing's paper from 1936 was in itself an excellent example of mathematical modeling. He first presents an analysis of how humans compute, i.e. he proposes a specific structure for computations. Then he creates a syntax, i.e. the abstract Turing machine, in order to establish the fundamental properties of computations. We recommend Gandy's paper where he tells the story of how the notion of computable function proposed by Turing became universally accepted, hence a "true" object of the mind of the species.

Much need still to be added to these few words about mathematical modeling as a general model. There is for example an important issue concerning explicit/conscious versus implicit/sub-conscious levels of both structure, syntax and algorithms. I have taken some notice of this in my reference to the cathedral builders of the middle ages above, but more needs to be said. At this point let me just stress the importance of the fact that all three ingredients, structure, syntax, algorithms, are abstract objects. A structure is an object of our culture which describes or pictures some part of the material and/or the mental, and even of the cultural world. Syntax and algorithms are tools created to understand and make use of structures. They are "obviously applicable". And if this is accepted, we are, in my view, a step closer to an understanding of "how we know".

This is the simple picture. But Wigner points to examples where some pre-existing theories, developed deep inside mathematics, have important applications. Today he would probably refer to the interaction between algebraic geometry, topology and quantum field theory, surely this "is something bordering on the mysterious". We are not so convinced. We believe that a first step to understand this mystery is to note that "pure" mathematics evolved out of applied mathematics *when syntax became structure*. The first example is, perhaps, the elementary theory of numbers. We speculated above on the neuronal and cultural origin of some rudimentary forms of arithmetic based on numbers, counting/addition and measurements/multiplication. But when this syntax for counting and measurement is itself seen as a new structure, we understand how the "rekurrirende Denkweise" and higher forms of mathematics could develop.

The transition from old syntax to new structure is not a feature restricted to the mathematical and natural sciences. In the case of language syntax was at first a tool to describe structures seen or experienced in the world. But syntax has also an internal structure, and this has inspired new mathematical theories, which in a sense is a syntax for a structural study of grammar. Similar remarks apply to computer science, in particular to the theory of programming languages. We could continue and almost everywhere across pure and applied mathematics uncover new examples of when old syntax becomes new structure. The resulting picture is complex with many interrelationships and movements back and forth between "structure" and "syntax". This is not a mystery, but presents a challenge. However involved the relationships may become, we still insist on the original analysis of mathematical modeling and the "obvious applicability" as a valid starting point. The challenge is to see how this applicability survives the many turns and twists of structure and syntax.

I have tried in this essay to strengthen the cultural view on the origin and growth of mathematics by making use of some recent results from evolutionary theory and neuroscience. Despite the philosopher's critique I believe that these results give added weight to the anthropologist's claim that mathematics is at the same time both dependent on and independent of the human mind, where "mind" in the first case is read as the "mind of the species" and in the second as the "individual mind". This could, perhaps, provide an *Ansatz* to a somewhat different view of the philosophy of mathematics than what is standard among philosophers. As the reader will have noticed, the work of Skolem has served as an inspiration for many parts of our discussion, in particular, his emphasis on structure and algorithms as prior to axioms and proofs.[79] However, despite the fact that the words "Begründung", "Grundlagenfragen", and "foundational research" occur frequently in the titles of his papers, he never worked out a complete and coherent view on the foundation of mathematics. I would like to believe that what I have here suggested, even if it is not a "true" Skolem philosophy of mathematics, is a view compatible with the many scattered remarks on foundational issues you can find in his *Selected Works*.[80]

We further note that our discussion is seen to strongly support a structural or semantic view not only of mathematics but of scientific methodology in general, different from the standard HD (hypothetic-deductive) approach. The HD methodology is, according to most textbook expositions, dominated by the sequence: hypothesis, deduction, test and possible falsification. This is a methodology dominated by syntax and logic. We have insisted on starting with what we "see", which in a sense is a semantic or model-theoretic alternative to the standard HD method, an approach early championed by P. Suppes.[81] In this approach structures and patterns are the foundation. We find a forceful expression of a similar view in Barrow[82]: "This reflection on the symmetries behind the laws of Nature also tells us why mathematics is so useful in practice. Mathematics is simply the catalogue of all possible patterns. Some of those patterns are especially attractive and are studied for decoration, others are patterns in time or in chains of logic. Some are described solely in abstract terms, while others can be drawn on paper or carved in stone. Viewed in this way, it is inevitable that the world is described by mathematics."

Let us in conclusion note, that if we stay within first order logic, as some philosophers seems to prefer, there is a perfect duality between syntax and semantics. If we move beyond this restricted logical frame, we recognize that the emphasis on structure and meaning provides us with a broader perspective.[83] I believe that the discussion of "what there is" and of "how we know" gives some evidence for this claim. If we, as we have insisted above, understand the family of

[79]Fenstad and Wang (2009), Sect. 4, and Cohen (2005).

[80]Skolem (1970).

[81]Suppes (2002).

[82]Barrow (2010).

[83]Consult Fenstad (2010) for a discussion of meaning in natural languages.

concepts—structure, syntax, algorithms—in a sufficiently broad sense, the discussion of the "obvious applicability" could in a natural way be extended from the sciences to the humanities and even to the arts.

9.4 Postscript: What Numbers Are

Our aim in the main part of this chapter was an attempt to understand the general nature of knowledge; this attempt, however, needs some additional remarks. Structures and algorithms are essential for understanding the nature of knowledge: what we see and how we can make use of what we see. And mathematics was claimed to be the link between the two. There is, however, no mathematics without numbers, either directly as objects or more generally as tools. Understanding knowledge thus means to know what numbers are.

Numbers are not objects in space and time; it thus seems necessary to acknowledge, as we did in the main part of the chapter, the existence of abstract objects in addition to physical objects and states of (individual) minds. But platonism in the sense of K. Gödel never seemed attractive to me; Husserl's phenomenology or a theory of "cultural space" (seen as the "collective mind of the species") could be possible alternatives. Husserl was strongly against a cultural point of view. I never was convinced by his line of reasoning and tried in the main part of the essay to develop the cultural approach. This discussion needs some additions. But first a general point.

Why is there philosophy of mathematics at all? This is the title of a recent book by Ian Hacking.[84] He is concerned with and reports extensively on actual mathematical praxis, on the nature of proofs and the variety of applications as seen within the community. His concerns are in a deep sense "foundational", but he appears to be highly skeptical of the usefulness of almost all there is of current philosophy of mathematics. There are, of course, "philosophical points" that need to be discussed, but rather than engaging with the standard philosophical discourse, Hacking's main references are to the mathematical community itself, names such as Lichnerowicz, Langlands and Gowers—all great names of modern mathematics—are rarely to be met in philosophy texts.

There are many points of contact between Hacking's book and my general approach. But there are differences, he prefers proofs, I subscribe to algorithms. This may seem as a minor difference in view of the close connection between proofs and computations. But the equal emphasis on structure and algorithms may mark a shift in our understanding of what mathematics should be, and how it is to be related to a general theory of knowledge—in this respect the von Neumann of numerical weather forecasting is a better guide than the von Neumann of axiomatic set theory.

I share Hacking's anti-Platonist stance, but still believe in abstract objects. The belief in mathematical objects is widespread within the mathematical community. I provided several references in the earlier essay, but I should also have included the

[84] Hacking (2014)

book *Why belief matters: reflections on the nature of science* by E. Brian Davies.[85] Davis is in no doubt about what mathematics is: "Mathematics is an aspect of human culture, just as are language, law, music, and architecture".[86] This is a strong and unqualified expression of how mathematics is grounded in human culture, a view that was severely criticized by Husserl: There are mathematical objects, but they are neither social or cultural objects, because such objects are bound to times and places, see the discussion of Husserl's philosophy of mathematics in Tieszen.[87] I tried in the main part of the essay to strengthen the cultural view against this criticism with arguments drawn both from neuroscience and evolutionary theory. These arguments seem to be strengthen by recent results which show, contrary to the common belief that evolution requires thousands or millions of years, that the human genome can change in as little as a generation, see the review.[88] Thus what is "seen" can leave an imprint on the human genome and hence become independent of times and places.

In this connection I want briefly to mention a forthcoming book,[89] *Thin objects: An Abstractionist Account*, by Ø. Linnebo as highly relevant to the approach developed in my text. His aim is to balance the need for an ontology of mathematical objects with an acceptable epistemology of the subject. We shall discuss his approach through an example. Elementary plane geometry has points and lines as its basic objects. We all recall that two lines are said to be parallel if they have no point in common. But to unravel the "deeper" properties of geometry, new constructions and objects seem necessary. We agree that parallel lines have the same direction, but are *directions* a new class of objects different from points and lines? We may argue that in speaking about directions we are just speaking about properties of lines, but for mathematicians engaged in constructing the field of coordinates of a given geometry, directions are "real" objects.[90] Linnebo develops in his book the idea how new objects can be introduced through suitable identity criteria and schemes of abstraction. In our case the direction of two lines is one and the same object if and only if they are parallel. Note that in this case the abstraction is predicative in the sense that it involves quantification only over what has already been accepted into our domain (namely lines).

We next turn to numbers. In foundational discussions one usually distinguishes between a *cardinal* versus an *ordinal* conception. The cardinal conception abstracts numbers from one-to-one mappings, and is much preferred by analytic philosophers. The ordinal conception takes counting and order as its basis. Linnebo argues, in my view quite convincingly, that the ordinal approach is the better one. The ordinal story also fits with the story as told in the main part of my text where I discussed the historical evidence proving how numbers grew out of 'bones, strings and standing stones'. Thus let a numeral be any concrete object that can stand in a suitable ordering,

---

[85] Davies (2010).

[86] Davies (2010), p. 10.

[87] Tieszen (1995).

[88] Pennisi (2016).

[89] See the draft Linnebo (2014).

[90] Artin (1957).

more precisely a *numeral* is a pair (u, R), where u is the concrete object and R is some ordering in which u occupies some position. In this brief summary we shall not discuss what properties to demand of the ordering R, except that it always should be a discrete linear ordering with an initial element. With each such pair (u, R) we associate a *number* N (u, R) subject to the identity criterion that two pairs (u, R) and (u', R') determine the same number if and only if u and u' occupy analogous positions in their respective orderings R and R'. The reader may test his understanding of this construction either through a simple drawing, or by writing down a precise definition using the notion of order-preserving correlations between initial segments.

The abstraction of numbers N (u, R) from concrete numerals (u, R) marks the beginning of arithmetic as a formal and abstract object of study. It is, however, interesting to note that there still exist well developed "numerical" skills within "kitchen and street-vendor" praxis entirely at the level of concrete numerals. Linnebo concludes his discussion of numbers and of thin objects by affirming that abstract mathematical objects exist, but we note that his anti-nominalism does not extend to a full mathematical platonism. He seems to argue that a positive theory of thin objects is sufficient to provide a satisfactory account of both mathematical ontology and mathematical epistemology. I have argued for a cultural view on the origin of numbers. This is a story compatible with the one Linnebo tells about thin objects. But more is needed if we want to understand how numbers, thus introduced, have become independent of times and places.

A second philosophy of arithmetic is the title of a recent paper[91] by Penelope Maddy. Her approach seem to have many points of contact with the views developed in my essay. "To summarize, then, the Second Philosopher concludes that much of the world displays a familiar abstract template – a domain, properties and relations, dependencies – amenable to a strong Kleene three-valued scheme; let me call this formal structuring KF, in honor of Kant and Frege, from whom it derives".[92] We may argue about the historical foundation, mathematicians may have other sources, but the view expresses a widespread attitude in the mathematical community: first structure, then syntax and rules. A forceful expression of this view is found in Rawls' report on the celebrated Sacks-Dreben Slugfest in 1993. Both are well-known Harvard professors, Dreben in philosophy, Sacks in mathematics. Sacks argued that what is important for logic is structure, not syntax. "In learning mathematics, we become aware of and experience structure. This is what mathematical experience is. … Sacks granted that syntax is simpler and that it therefore makes sense that we begin teaching logic with syntax".[93]

Maddy jumps immediately to Kleene's three-valued logic. This is appropriate for simple partial models, and it is also a syntax/logic well adapted to situation semantics.[94] But what we "see" may dictate different choices.

---

[91]Maddy (2014).

[92]Maddy (2014), p. 225.

[93]Rawls (2001), p. 420.

[94]Fenstad et al. (1987).

The Second Philosopher and I seem to a large extent to agree on the question of applicability. It is important to keep in mind that the "mind of the species", where structures and algorithms live, only interacts with nature, where measuring instruments and computing devices live, through an individual mind. This is a valid general observation, but not in any sense an answer to how for example abstract computational structures and physical calculations on a piece of paper are connected. We know of one case where the neural substructure of a cognitive task is beginning to be understood, the case of the so-called grid and (the later discovered) speed cells.[95] These cells and the geometrical structure they display form a neuron-based navigation system that can tell animals where they are going and where they have been. Recently we have seen a first attempt to extend this kind of approach to an analysis of how the brain organizes conceptual knowledge.[96] But beyond this there are few hard facts.[97]

Our interest is how to connect abstract computational structures and physical calculations. Let us first make a few remarks on language and brain.[98] In this case we assume that what we "see"—objects and properties—have through an evolutionary process given rise to some rudimentary form of "universal" proto-syntax (such as the decomposition of a whole message S into an NP + VP structure), which in turn has left an imprint on the human genome. We would like to believe that the situation with numbers and calculations is similar to the language case. As noted above the speed of evolutionary processes seems to be much faster than we originally believed, from millions of years we now move to generations or at most some thousand years. This makes it possible that the abstraction of numbers as cultural objects from rudimentary computational praxis as described above, has in turn left an imprint on the human genome. But what is imprinted is still largely unknown; see, however, the discussion in Cappelletti and Giardino[99] and the further comments made in Remark 2 of Chap. 8 on the changed role of biological memory. The exact neural mechanisms are still unknown, but perhaps some clues can be found in the earlier models of Dehaene and Changeux.[100] If this is a correct line of reasoning, we can argue contra Husserl that numbers, invented by man, have become independent of times and places.

In some sense there also seems to be an implicit cultural foundation for Maddy's Second Philosophy. My approach is more directly cultural, and believing in numbers as (abstract) objects, I am probably a "thin-realist" in her view. Maddy seem to believe that we have a free choice between being an "arealist" or a "thin realist". However, no one who has read C. P. Snow[101] on Hardy's visit to Ramanujan in hospital will

[95] See Hafting et al. (2005), Kropff et al. (2015) and also the survey article Abbott (2014).

[96] Constantinescu et al. (2016).

[97] Marcus et al. (2014) gives an update to our current understanding of how to bridge the gap from neuronal networks to symbolic cognition.

[98] Fenstad (2010).

[99] Cappelletti and Giardino (2007).

[100] Dehaene and Changeux (1993), see also the updates in Dehaene and Brannon (2010) and Dehaene (2014).

[101] Snow (1967).

doubt that numbers exist: "Hardy had gone out to Putney by taxi, his usual method of conveyance. He went into the room where Ramanujan was lying. Hardy, always inept about introducing a conversation, said, probably without a greeting, and certainly as his first remark: 'I thought the number of my taxi was 1729. It seemed to me a rather dull number.' To which Ramanujan replied: 'No, Hardy! It is a very interesting number. It is the smallest number expressible as the sum of two cubes in two different ways.'"

There are two lessons to be learned. First, numbers exist. Second, there is a difference between knowing a number and knowing what properties a number has. In this respect abstract and physical objects are similar, "invention" and "discovery" are not necessarily contradictory. Numbers are cultural objects invented by man. But no man knows what properties there are to be discovered.

What little we know was the concluding point of an earlier book *Grammar, Geometry, and Brain*[102] Many "facts" are still unknown, the gap between the mind and the brain still remains, but I have one firm belief, that a proper understanding of cultural space—the collective mind of the species—will provide a sound foundation for an analysis of abstract objects such as number, mind and meaning. Cultural space is not only necessary but also sufficient for general knowledge.[103]

References

Abbott, A. (2014). Brains of Norway, Nobel Prizewinners May-Britt Moser and Edvard Moser have spent a career together near the Arctic Circle exploring how our brains know where we are. *Nature*, 514, 154–157.

Albeverio, S., Fenstad, J. E., Høegh-Krohn, R., & Lindstrøm, T. (1986). *Nonstandard methods in stochastic analysis and mathematical physics*. New York: Academic Press. (Reprinted by Dover Press 2009).

Amit, D. J. (1989). *Modeling brain function*. Cambridge, UK: Cambridge University Press.

Artin, E. (1957). *Geometric algebra*. New York: Interscience Publisher.

Atiyah, M., et al. (2006). The interface between mathematics and physics. *Irish Mathematical Society Bulletin*, 58, 33–54.

Barrow, J. D. (2010). Simple really: From simplicity to complexity—and back again. In B. Bryson (Ed.), *Seeing further, the story of science and the Royal Society*. London: Harper Press.

Barwise, J. (1985). In Barwise & Feferman (Eds.), *Model-theoretic logics: Background and aims* (pp. 4–5).

Barwise, J., & Feferman, S. (Eds.). (1985). *Model-theoretic logics*. Heidelberg: Springer.

Beller, S., & Bender, A. (2008). The limits of counting: Numerical cognition between evolution and culture. *Science, 319*, 213–215.

Bergen, B. (2012). *Louder than words: The new science of how the mind makes meaning*. New York: Basic Books.

Cappelletti, M., & Giardino, V. (2007). In M. Leng et al. (Eds.), *The cognitive basis of mathematical knowledge* (pp. 74–83).

[102]Fenstad (2010).

[103]It is with great pleasure that I thank Johan van Benthem, Sol Feferman, Dagfinn Føllesdal, Mikhail Katz, Øystein Linnebo, Richard Tieszen, Herman Ruge Jervell, Dana Scott, Patrick Suppes and Ed Zalta for many helpful and critical comments on several earlier drafts of this paper.

Capra, F. (2007). *The science of Leonardo*. New York: Anchor Books.

Cohen, P. J. (2005). Skolem and pessimism about proof in mathematics. *Philosophical Transactions of the Royal Society of London A: Mathematical, Physical and Engineering Sciences, 363*, 2407–2418.

Cole, J. (2008). Mathematical domains: Social constructs? In B. Gold & R. A. Simon (Eds.), *Proof and other dilemmas: Mathematics and philosophy*. Washington, D.C.: Mathematical Association of America.

Cole, J. (2012). Toward an institutional account of the objectivity. *Necessity, and Atemporality of Mathematics, Philosophia Mathematica, 21*, 9–36.

Constantinescu, A. O., O'Reilly, J. X., & Behrens, T. E. J. (2016). Organizing conceptual knowledge in humans with a gridlike code. *Science, 353*, 1464–1468.

Davies, E. B. (2010). *Why beliefs matter reflections on the nature of science*. Oxford: Oxford University Press.

Davies, P. J., & Hersh, R. (1980). *The mathematical experience*. Boston: Birkhäuser.

Dehaene, S. (1997). *The number sense: How the mind creates mathematics*. London: The Penguin Press.

Dehaene, S. (2014). *Consciousness and the brain, deciphering how the brain codes our thoughts*. New York: Viking Press.

Dehaene, S., & Brannon, E. M. (2010). Space. *Time, and Number: A Kantian Research Program, Trends in Cognitive Sciences, 14*, 517–519.

Dehaene, S., & Changeux, J. P. (1993). Development of elementary numerical abilities: A neuronal model. *Journal of Cognitive Neuroscience, 5*, 390–407.

Donald, M. (1990). *Origin of the modern mind: Three stages in the evolution of culture and cognition*. Cambridge, USA: Harvard University Press.

Donald, M., et al. (1993). Precis and discussion of origin of modern mind. *Behavioral and Brain Science*, 16(4), 737–748.

Dunn, M., Greenhill, S. J., Levinson, S. C., & Gray, R. D. (2011). Evolved structure of language shows lineage-specific trends in word-order universals. *Nature, 473*(7345), 79.

Ehrlich, P. (2006). The rise of non-Archimedian mathematics and the roots of a misconception. I. The emergence of non-Archimedian systems of magnitudes. *Archive for History of Exact Sciences, 60*, 1–121.

Ehrlich, P. (2012). The absolute arithmetic continuum and the unification of all numbers great and small. *The Bulletin of Symbolic Logic, 18*, 1–45.

Eloy, C. (2011). Leonardo's rule, self-similarity and wind-induced stresses in trees. *Physical Review Letters, 107*, 258101–258105.

Feferman, S., et al. (1990). *Kurt Gödel. Collected works, vol II*. New York and Oxford: Oxford University Press.

Feferman, S., et al. (1995). *Kurt Gödel. Collected works, vol III*, New York and Oxford: Oxford University Press.

Feferman, S., et al. (2003). *Kurt Gödel. Collected works, vol V*, New York and Oxford: Oxford University Press.

Feferman, S. (2009). Conceptions of the continuum. *Intellectica, 51*, 169–189.

Fenstad, J. E. (1985). Is nonstandard analysis relevant for the philosophy of mathematics? *Synthese, 62*, 289–301.

Fenstad, J. E. (1987). The discrete and continuous in mathematics and the natural sciences. In G. Toraldo di Francia (Ed.), *Infinities in Science* (pp. 111–125). Rome: Istituto della Enciclopedia Italiana.

Fenstad, J. E. (1990). In Feferman et al. *Introductory note to Gödel 1974* (pp. 307–310).

Fenstad, J. E. (2010). *Grammar, geometry, and brain*. Chicago: University of Chicago Press.

Fenstad, J. E., & Wang, H. (2009). Thoralf Albert Skolem. In D. M. Gabbay & J. Woods (Eds.), *Handbook of the History of Logic* (Vol. 5, pp. 127–194). Amsterdam: North-Holland.

Fenstad, J. E., Halvorsen, P. K., Langholm, T., & van Benthem, J. (1987). *Situations, language and logic*. Dordrecht: D. Reidel Publishing Company.

Ferguson, E. S. (1993). *Engineering and the mind's eye*. Cambridge, MA: MIT Press.
Fisher, S. E., & Ridley, M. (2013). Culture, genes, and the human revolution. *Science, 340,* 929–930.
Føllesdal, D. (1995a). Gödel and Husserl. In J. Hintikka (Ed.), *From Dedekind to Gödel: Essays on the development of the foundation of mathematics* (pp. 427–446). Dordrecht: Kluwer.
Føllesdal, D. (1995b). In Feferman et al. (Eds.), *Introductory note to Gödel 1961* (pp. 364–373).
Fraenkel, A. (1928). *Einleitung in die Mengenlehre* (3rd ed.). Berlin: Springer.
Gandy, R. (1988). In Herken (Ed.), *The confluence of ideas in 1936* (pp. 55–111).
Gärdenfors, P. (2000). *Conceptual spaces, the geometry of thought*. Cambridge, MA: MIT Press.
Gödel, K. (1961). In Feferman (Ed.), *The modern development of the foundations of mathematics in the light of philosophy* (pp. 374–387).
Gödel, K. (1974). In Feferman et al. (Eds.), *Remark on non-standard analysis* (p. 311).
Gödel, K. (2003). In Feferman et al. (Eds.), *Correspondence with Abraham Robinson* (pp. 195–206).
Gordon, P. (2004). Numerical cognition without words: Evidence from Amazonia. *Science, 306,* 496–499.
Grattan-Guinnes, I. (2004). The mathematics of the past: Distinguishing its history from our heritage. *Historia Mathematica, 31,* 163–185.
Hacking, I. (2014). *Why is there philosophy of mathematics at all?*. Cambridge: Cambridge University Press.
Hafting, T., Fyhn, M., Molden, S., Moser, M. B., & Moser, E. (2005). Microstructure of a spatial map in the entorhinal cortex. *Nature, 436,* 801–806.
Hauser, K. (2006). Gödel's program revisited. Part I: The turn to phenomenology. *The Bulletin of Symbolic Logic, 12,* 529–590.
Herken, R. (1988). *The universal turing machine. A half-century survey*. Oxford: Oxford University Press.
Hersh, R. (1997). *What is mathematics, really?*. Oxford: Oxford University Press.
Joseph, G. G. (1991). *The crest of the peacock, non-European roots of mathematics*. London: Penguin Books.
Kanovei, V., Katz, M., & Mormann, T. (2013). Tools. *Objects, and Chimeras: Connes on the Role of Hyperreals in Mathematics, Foundations of Science, 18,* 259–296.
Katz, M. G., & Sherry, D. (2013). Leibniz's infinitesimals: Their fictionality, their modern implementations, and their foes from Berkeley to Russel and beyond. *Erkenntnis, 78,* 571–625.
Keisler, H. J. (1994). The hyperreal line. In P. Ehrlich (Ed.), *Real numbers, generalizations of the reals, and theories of continua* (pp. 207–237). Dordrecht: Kluwer Academic Publishers.
Keisler, H. J. (2007). In van den Berg & Neves (Eds.), *The strength of nonstandard analysis*.
Kitcher, P. (1984). *The nature of mathematical knowledge*. Oxford: Oxford University Press.
Klein, S. (2010). *Leonardo's legacy*. Cambridge Mass: Da Capo Press.
Kropff, E., Carmichael, J. E., Moser, M. B., & Moser, E. (2015). Speed cells in the medial entorhinal cortex. *Nature*, published online July 15, 2015.
Laland, K. N., Odling-Smee, J., & Myles, S. (2010). How culture shaped the human genome: Bringing genetics and the human sciences together. *Nature Reviews/Genetics, 11,* 137–148.
Leng, M. (2007). In Leng et al. (Eds.), *Introduction* (pp. 1–15).
Leng, M., Paseau, A., & Porter, M. (Eds.). (2007). *Mathematical knowledge*. Oxford: Oxford University Press.
Linnebo, Ø. (2008). The nature of mathematical objects. In B. Gold & R. A. Simon (Eds.), *Proof and other dilemmas: Mathematics and philosophy*. Washington, D.C.: Mathematical Association of America.
Linnebo, Ø. (2009). The individuation of the natural numbers. In O. Bueno & Ø. Linnebo (Eds.), *New waves in philosophy of mathematics*. London: Palgrave.
Linnebo, Ø. (2014). *Thin objects: An abstractionist account*. Oslo: University of Oslo.
Machover, M. (2003). In Feferman et al. (Eds.), *Abraham Robinson: Introductory note* (pp. 191–194).
Maddy, P. (2014). A second philosophy of arithmetic. *The Review of Symbolic Logic, 7,* 222–249.

Marcus, G., Marblestone, A., & Dean, T. (2014). The atoms of neural computations. *Science, 346,* 551–552.
Mumford, D. (2000). The dawning age of stochasticity. In Arnold, Atiyah, Lax, & Mazur (Eds.), *Mathematics: Frontiers and perspectives* (pp. 197–218). Providence: American Math Society.
Mumford, D. (2002), Pattern theory: The mathematics of perception. In *Proceedings of ICM 2002* (Vol. I, pp. 401–422). Beijing.
Newman, J. (1956). *The world of mathematics*. Simon and Schuster: New York.
Nodelman, U., & Zalta, E. (2016). *Foundations for mathematical structuralism*. Stanford: Center for the Study of Language and Information.
Normile, D. (2012). Experiments probe language's origin and development. *Science, 336,* 408–411.
Nowak, M. A., Plotkin, J. B., & Jansen, V. A. A. (2000). The evolution of syntactic communication. *Nature, 404,* 495–498.
Pennisi, E. (2016). Tracking how humans evolve in real time. *Science, 352,* 876–877.
Popper, K. (1972). *Objective knowledge. An evolutionary approach*. Oxford: The Clarendon Press.
Rawls, J. (2001). Afterword: A reminiscence. In J. Ford & S. Shieh (Eds.), *Future pasts. The analytic tradition in twentieth-century philosophy*. Oxford: Oxford University Press.
Robinson, A. (1965). Formalism 64. *Logic, Methodology, and Philosophy of Science*. Amsterdam: North-Holland.
Robinson, A. (1966). *Nonstandard analysis*. Amsterdam: North-Holland.
Rosch, E. (1983). Prototype classification and logical classification. In E. Scholnik (Ed.), *New trends in cognitive representations*. Hillsdale, NJ: Lawrence Erlbaum Ass.
Skolem, T. A. (1917). *Untersuchungen über einige Klassen kombinatorischer Probleme* (No. 6, pp. 1–99). Oslo: K.V. Skr. I.
Skolem, T. A. (1920). *Logisch-kombinatorische Untersuchungen über die Erfüllbarkeit und Beweisbarkeit mathematischen Sätze nebst einem Theoreme über dichte Mengen* (No. 4, pp. 1–36). Oslo: K. V. Skr. I. (Reprinted in Skolem 1970, pp. 103–136. English translation in van Heijenoort 1967, pp. 252–263).
Skolem, T. A. (1923). *Begründung der elementären Arithmetik durch die rekurrierende Denkweise ohne Anwendung scheinbarer Veränderlichen mit unendlichem Ausdehnungsbereich* (No. 6, 38 pp). Oslo: K.V.Skr. I. (Reprinted in Skolem 1970, pp. 153–188. English translation in van Heijenoort 1967, pp. 302–333).
Skolem, T. A. (1928). Über die mathematische Logik. *Norsk Matematisk Tidsskrift, 10,* 125–142. (Reprinted in Skolem 1970, pp. 189-206. English translation in van Heijenoort 1967, pp. 508–524).
Skolem, T. A. (1929). Über die Grundlagendiskussionen in der Mathematik. In *Proceedings of the 7th Scandinavian Mathematical Congress* (pp. 3–21), Oslo. (Reprinted in Skolem 1970, pp. 207–225).
Skolem, T. A. (1933). *Undersøkelser over potensrester og over logisk karakterisering av tallrekken* [Investigations on power residues and on logical characterisation of the number series] (Vol. III. no 4, 25 pp). Bergen: Christian Michelsens Institutt.
Skolem, T. A. (1934). Über die Nichtcharakterisierbarkeit der Zahlenreihe mittels endlich oder abzählbar unendlich vieler Aussagen mit ausschliesslich Zahlenvariablen. *Fundamenta mathematicae, XXIII,* 150–161. (Reprinted in Skolem 1970, pp. 355–366).
Skolem, T. A. (1970). In J. E. Fenstad (Ed.), *Selected works in logic*. Oslo: Universitetsforlaget.
Smith, B., & Smith, D. W. (1995). *The Cambridge companion to Husserl*. Cambridge: Cambridge University Press.
Snow, C. P. (1967). Foreword to G. H. Hardy. *A mathematician's apology*. Cambridge: Cambridge University Press.
Suppes, P. (2002). *Representation and invariance of scientific structures*. Stanford: CSLI.
Suppes, P. (2011). Why the effectiveness of mathematics in the natural sciences is not surprising. *Interdisciplinary Science Reviews, 36,* 244–254.
Tadmor, E. (2012). A review of numerical methods for nonlinear partial differential equations. *Bulletin AMS, 49,* 507–554.

Tieszen, R. (1995). In Smith et al. (Eds.), *Mathematics* (pp. 438–462).
Tieszen, R. (2010). Mathematical problem-solving and ontology: An exercise. *Axiomathesis, 20,* 295–312.
Tieszen, R. (2011). *After Gödel: Platonism and rationalism in mathematics and logic*. Oxford: Oxford University Press.
Turing, A. M. (1936). On computable numbers, with an application to the Entscheidungsproblem. *Proceedings of the London Mathematical Society, 42,* 230–265.
van den Berg, I. P., & Neves, V. (Eds.). (2007). *The strength of nonstandard analysis*. Wien and New York: Springer.
van Heijenoort, J. (1967). *From Frege to Gödel*. Cambridge, MA: Harvard University Press.
von Neumann, J. (1947). The mathematician. In R. B. Heywood (Ed.), *The works of the mind* (pp. 180–196). Chicago: University of Chicago Press.
White, L. A. (1947). The locus of mathematical reality. *Philosophy of Science, 14*(4), 289–303. (Reprinted in Newman 1956, pp. 2348–2364).
Whitehead, A. N., & Russell, B. (1910–1913). *Principia Mathematica* (3 vol). Cambridge: Cambridge University Press.
Wigner, E. P. (1960). The unreasonable effectiveness of mathematics in the natural sciences. *Communications in Pure and Applied Mathematics, 13*(1), 1–14. Reprinted in Wigner (1979). *Symmetries and Reflections*, (pp. 222–237), Woodbridge, CT: Ox Bow Press.

The manufacturer's authorised representative in the EU is Springer Nature Customer Service Centre GmbH, Europaplatz 3, 69115 Heidelberg, Germany. If you have any concerns regarding our products, please contact ProductSafety@springernature.com

Printed and bound by CPI Group (UK) Ltd, Croydon, CR0 4YY
30/09/2026
02225606-0014